I0488592

Special Publication 500-295

Conformance Testing Methodology for ANSI/NIST-ITL 1-2011, Data Format for the Interchange of Fingerprint, Facial & Other Biometric Information (Release 1.0)

NIST/ITL Conformance Testing Methodology Working Group

Fernando L. Podio
Dylan Yaga
Christofer J. McGinnis
Editors

http://dx.doi.org/10.6028/NIST.SP.500-295

**National Institute of
Standards and Technology**
U.S. Department of Commerce

Special Publication 500-295

Conformance Testing Methodology for ANSI/NIST-ITL 1-2011, Data Format for the Interchange of Fingerprint, Facial & Other Biometric Information (Release 1.0)

NIST/ITL Conformance Testing Methodology Working Group

Fernando L. Podio
Dylan Yaga
Christofer J. McGinnis
Editors
Computer Security Division
Information Technology Laboratory

http://dx.doi.org/10.6028/NIST.SP.500-295

August 2012

U.S. Department of Commerce
Acting Secretary Rebecca M. Blank

National Institute of Standards and Technology
Patrick Gallagher, Under Secretary for Standards and Technology and Director

Reports on Information Technology

The Information Technology Laboratory (ITL) at the National Institute of Standards and Technology (NIST) stimulates U.S. economic growth and industrial competitiveness through technical leadership and collaborative research in critical infrastructure technology, including tests, test methods, reference data, and forward-looking standards, to advance the development and productive use of information technology. To overcome barriers to usability, scalability, interoperability, and security in information systems and networks, ITL programs focus on a broad range of networking, security, and advanced information technologies, as well as the mathematical, statistical, and computational sciences. Special Publication 500-series reports on ITL's research in tests and test methods for information technology, and its collaborative activities with industry, government and academic organizations.

Certain commercial entities, equipment, or materials may be identified in this document in order to describe an experimental procedure or concept adequately. Such identification is not intended to imply recommendation or endorsement by the National Institute of Standards and Technology, nor is it intended to imply that the entities, materials, or equipment are necessarily the best available for the purpose.

National Institute of Standards and Technology
Special Publication 500-295
Natl. Inst. Stand. Technol.
Spec. Pub. 500-295
204 pages

Foreword

The existence of biometric standards alone is not enough to demonstrate that products meet the technical requirements specified in the standards. Conformance testing captures the technical description of a specification and measures whether an implementation faithfully implements the specification. Conformance testing provides developers, users, and purchasers with increased levels of confidence in product quality and increases the probability of successful interoperability.

Although no conformance test can be comprehensive enough to test all the different combinations of mandatory requirements of a standard and all possible combinations of conditional and optional characteristics that could be included in ANSI/NIST-ITL 2011 (AN-2011) transactions, a well-designed conformance test tool that faithfully implements a standard conformance testing methodology could raise the level of confidence on the test results. Therefore, transactions tested with such a tool (and reported to be conformant to the standard), are more likely conform to the standard.

Introduction

This first edition of the Conformance Testing Methodology (CTM) for AN-2011 includes comprehensive tables of AN-2011 requirements and assertions for the selected Record Types supported by this edition. The tables of requirements and assertions indicate which assertions apply to the Traditional encoding format, which apply to the National Information Exchange Model (NIEM)-compliant encoding format, and which apply to both encoding formats. A test assertion syntax is specified to clearly define the assertions associated with each requirement. Some requirements are not supported by the testing methodology, and are identified as exceptions. When an exception is present, the requirement is listed in the tables, but no assertions are defined for that requirement. Justification is provided for each exception.

Table of Contents

List of Tables and Figures

1 Scope

This conformance testing methodology includes the concepts and test types necessary to test transactions for conformance to the AN-2011 standard. It defines three levels of conformance testing: Level 1, Level 2, and Level 3 conformance testing (see Sec. 5.3). This edition of the conformance testing methodology specifies requirements and test assertions for the following sections and Record Types of the AN-2011 standard:

- Section 5: Data Conventions
- Section 7: Information Common to Several Record Types
- Section 8.1 Record Type-1: Transaction information record
- Section 8.4 Record Type-4: Grayscale fingerprint image
- Section 8.10 Record Type-10: Facial, other body part and SMT image record
- Section 8.13 Record Type-13: Friction-ridge latent image record
- Section 8.14 Record Type-14: Fingerprint image record
- Section 8.15 Record Type-15: Palm print image record
- Section 8.17 Record Type-17: Iris image record
- Annex B: Traditional Encoding

Requirements and assertions for the deprecated Record Types 3, 5, and 6 are also included to ensure nonexistence of the deprecated records. Additionally, an assertion is specified that checks for the nonexistence of reserved Record Types 11, 12, and 22 through 97. A complete description of support for Record Types and interrelated fields is provided in Annex A. Tables of terms, operands, and operators that are used in specifying the test assertions are included. The assertion tables identify requirements for Level 1 and 2 conformance testing. In addition, the assertion tables identify requirements for Level 3 only when it is necessary to clarify that the requirement is categorized as Level 3. Only test assertions for Level 1 and Level 2 conformance testing are part of the scope of this conformance testing methodology. Implementation exceptions are identified in Sec. 6.4. Type-A and Type-B conformance testing are defined. Only Type-A testing is specified. Type-B conformance testing is outside of the scope of this edition of the CTM.

This conformance testing methodology does not establish tests of characteristics (i.e., performance, acceptance, security, robustness) of products that generate the transactions.

2 Conformance

AN-2011 conformance test tools that claim conformance to this conformance testing methodology shall satisfy the requirements of the testing methodology specified in Sec. 5 and shall follow the procedures defined by Level 1 and Level 2 test assertions as specified in Sec. 6.

3 Normative references

NIST Special Publication 500-290, ANSI/NIST-ITL 1-2011, November 2011, *Information Technology:* American National Standard for Information Systems - Data Format for the Interchange of Fingerprint, Facial & Other Biometric Information

JPEG (Joint Photographic Experts Group), *JPEG File Interchange Format, Version 1.02.* Available at http://www.jpeg.org/public/jfif.pdf

ISO/IEC 15444-1, *JPEG 2000, Information Technology - Digital Compression and Coding of Continuous-Tone Still Images Part 1: Requirements and Guidelines.*

ISO/IEC 15444-2, *Information technology — JPEG 2000 image coding system: Extension, available at:* http://www.jpeg.org/metadata/15444-2.PDF

ISO/IEC 15948:2004 *Information Technology -- Computer graphics and image processing -- Portable Network Graphics (PNG): Functional specification.*

IAFIS-IC-0110 (V3.1) *WSQ Gray-scale Fingerprint Image Compression Specification,* October 4, 2010.

4 Terms and definitions

assertion

A test procedure that represents a specific aspect of a requirement found in the base standard. The assertion is expressed using the operator and operand syntax defined by the conformance testing methodology.

base standard

ANSI/NIST-ITL 1-2011, Data Format for the Interchange of Fingerprint, Facial & Other Biometric Information, NIST Special Publication 500-290.

conformance

The adherence of an implementation to all specified requirements as defined in the base standard.

CTA

Conformance Testing Architecture.

CTM

Conformance Testing Methodology.

CTS

Conformance Testing Suite.

implementation

An ANSI-NIST-ITL 1-2011 transaction.

IUT

Implementation under test. The implementation supplied by a vendor to a laboratory for conformance testing.

test

Also known as a conformance test or assertion test, it is the execution of the testing procedure defined by an assertion or set of assertions in order to obtain a statement of conformance. The result of the test is a Boolean value that determines the implementation's conformity for the assertion. For a given requirement, if all tests pass for the associated assertions, then the implementation is considered to be conformant for that requirement.

Type-A testing

Type-A conformance testing checks the conformance of AN-2011 transactions to the requirements in the base standard.

Type-B testing

Type-B testing checks the ability to use an AN-2011, for example in a software application.

5 Conformance testing methodology

The conformance testing methodology defined within this publication addresses only Level 1 and 2 testing and Type-A testing. While some Level-3 requirements are identified in the tables of requirements and assertions, testing methodologies are provided for Level-1 and Level-2 conformance testing only (see "Hierarchy of conformance tests" for information regarding the three levels of conformance testing). Type-B testing is not defined.

5.1 Functional documentation of requirements

The tables of requirements and assertions included in this publication identify requirements defined in the base standard and provide a succinct listing of the information necessary to facilitate the development of conformance testing tools. Extracting the requirements from the base standard improves the efficiency of conformance test tool development by providing the information required in a consolidated manner.

3

Each requirement identified in this publication is associated with one or more assertions which collectively form the complete set of test procedures required to test an implementation for conformance to that requirement.

5.2   Limitations and exceptions

While conformance of an implementation to relevant requirements can be determined (e.g., content and relationships between elements), no test tool is guaranteed to be comprehensive and prove that a given system generating or using AN-2011 transactions is conformant under all possible circumstances. Well-designed conformance tests can, however, test the most likely sources of problems and demonstrate non-conformity (i.e., if errors are found, non-conformance of the transaction shall be proven), but the absence of detected errors does not necessarily imply full conformance to the standard.

The tables of requirements and assertions define all normative requirements as well as optional and conditional (dependent) requirements for selected Record Types found in the base standard (AN-2011). Some of the AN-2011 requirements in the tables do not have an associated assertion or set of related assertions due to the fact that some conformance tests require additional research, or that the test of that requirement is not feasible at the present time.

5.3   Hierarchy of conformance tests

Three levels of conformance testing are defined below. See also expanded discussions on these levels of conformance tests in Section 2 of NIST Special Publication 500-290. For each assertion included in the tables of requirements and assertions, a level of conformance testing is indicated.

Level 1 conformance testing

Level 1 conformance testing deals with the form and structure of the internal content and verifies that data structures exist and have allowable values. Specifically, it checks for the presence, structure, and value of each field, subfield, and information item in a transaction for conformance with the specification of the standard, both in terms of ranges and cardinality. Since Level 1 testing can be performed by a simple field-by-field reading of the standard and comparison to known values and their encoding, only the AN-2011 transactions are required for conformance testing, and not any hardware or software components used to create those transactions.

Level 2 conformance testing

Level 2 conformance testing deals with explicit requirements that check for internal consistency. Specifically, morphological conformance checks the relationships between fields, subfields, or information items within a transaction, including comparisons of values, as specified in the AN-2011 standard. Level 2 tests involve interactions between multiple values from different parts of the standard and sometimes from implicit observations that are not explicitly stated in the base standard. Thus, Level 2 tests require more complex validation than Level 1. Similar to Level 1 testing, Level 2 conformance testing only requires an AN-2011 transaction(s).

Level 3 conformance testing

Level 3 conformance testing checks if the biometric transaction is a faithful representation of the parent biometric data and ensures requirements are satisfied that are not merely Level 1 and Level 2 tests. Individual fields may have explicit semantic requirements for which conformance testing is significantly difficult or even impossible to test. Unlike Level 1 and Level 2 testing, Level 3 testing may require software and hardware components used to create the AN-2011 transactions, and may also require the subject and samples from which the biometric information stored in the transaction was collected. The requirements and assertion tables indicate whether Level 1 or Level 2 conformance testing is required to address the assertion identified in the test assertion. Required Level 3 conformance tests are not performed but they are identified in the tables to indicate that the requirement is not addressed or that it is not currently testable.

5.4 Conformance statements

In addition to providing a succinct list of the requirements in the AN-2011 standard, the tables of requirements and assertions provide the means for the developers of implementations under test (IUT) to claim in the tables the list of all the requirements supported including:
- Transaction information
- Encoding requirements
- Each Record Type included in the Transaction
  - Fields
  - Subfields
  - Information Items

This information is useful to the IUT supplier as a checklist on the content of their implementations and also useful to testing laboratories that would evaluate conformance of these IUTs against the supplier's claims. Two columns in the tables are included to provide this information: Implementation Support column (YES/NO/Partial) and Supported Range column (if Implementation Support is "Partial", the supported range should be provided).

If the IUTs are sent to a testing laboratory, the IUT provider shall also submit the information below to the laboratory:

- Provider name
- Provider address
- Transaction identifier
- Transaction version number
- Additional implementation information (optional)
- Submission date
- For each claimed Record Type, provide the Record Type number and whether or not (Yes or No) there are any known deviations from (or exceptions to) the requirements found in the base standard and identified in the Conformance Testing Methodology for

the associated Record Types in the IUT. For specific exceptions, the Implementation Support column of the tables of requirements and assertions shall be used to indicate the difference on a per-assertion basis. In addition, if the deviation is general and applies to the entire Record Type, a description shall be provided. This option is useful for cases where there have been modifications to the base standard that are not reflected in the conformance testing methodology, where the IUT provider believes there is a defect in the base standard or conformance testing methodology, and other instances where the implementation does not fully conform to the AN-2011 standard requirements.

The testing laboratory may use testing tools that implement or conform to this conformance testing methodology to provide a determination of the level of conformance of the IUT to the AN-2011 standard.

5.5 Test Assertion Syntax

Test assertions are expressed according to the operators and operands found in the tables of Operator Definitions and Operand Definitions, except for those instances where the assertion cannot be clearly or easily represented in a mathematical format. In those cases, English is used to express the assertion, and the text is contained within the < > characters.

5.5.1 Operators

The table below includes a complete description of the operators used throughout the requirements and assertion tables.

Table 5.1 - Assertion Syntax: Operator Definitions

Operator	Name	Description
	Operator Definitions	
AND	Logical And	Tests if both values are true.
ELSE	Else	Combined with the IF operator to specify what expressions are evaluated when the IF expression is false.
EQ	Equal To	Tests for equality between two values.
GT	Greater Than	Tests if the first value is greater than the second value
GTE	Greater Than or Equal To	Tests if the first value is greater than or equal to the second value.
IF	Logical If	Determines if the value or expression is true or false.
IFF	IF and Only IF	Tests the bi-conditional where each of the first and second expressions implies the other.
in	Container Specification	For X in Y, selects only those X found in Y.
LT	Less Than	Tests if the first value is less than the second value.
LTE	Less Than or Equal To	Tests if the first value is less than or equal to the second value.
MO	Member Of	Tests if the value is a contained within the set.
MOD	Modulo	For X MOD Y, provides the remainder of X divided by Y.
NEQ	Not Equal To	Tests for non-equality between two values.
NOT	Negate	Negates any operator or expression that follows.
OR	Logical Or	Tests if either value is true

P:N in Q	Query	Selects the Nth occurrence of P in Q.
ST	Such That	Enforces a condition upon the specified value or expression.
THEN	Then	Combined with the IF operator to specify what expressions are evaluated when the IF expression is true.
to	Range Selection	For X to Y, selects a set of values Z ST Z GTE X AND LTE Y
#	All	Provides all valid values.
:	Data Element Selection	For X:N, selects the Nth element in X.
,	Range Concatenation	For X,Y, represents the set of values containing both X and Y.
.	Field Selection	For X.Y, selects the field specified by Y in Record X.
<>	English Expression	Contains English text that could not be reasonably expressed mathematically.
{}	Value	For {X}, provides the value of X.
[]	Set	The set to be tested.

5.5.2   Terms

The table below provides a complete description of the terms used throughout the requirements and assertion tables.

Table 5.2 - Assertion Syntax: Terms

Term Definitions		
Term	**Name**	**Description**
Field(s)	Field	Field structure as defined by the AN 1-2011 standard.
InfoItem	US Separated Information Item	Information Item separated by the ASCII US (0x1F) separator character
Integers	Integer Set	Set of all integers.
NA	Not Applicable	The test or condition is not applicable.
Unsupported	Unsupported	The requirement is not supported in this version of the CTM. This may be the result of the related conformance test requiring additional research, or the result of the test being infeasible (level 3 only).
Record(s)	Record	Record structure as defined by the AN 1-2011 standard.
Subfield	RS Separated Subfield	Subfield separated by the ASCII RS (0x1E) separator character
Transaction	Transaction	Transaction structure as defined by the AN 1-2011 standard.
TRUE	True	The test always evaluates to true because there is no defined value for testing, or there is no value for which the test will fail.

5.5.3   Operands

The table below includes a complete description of the operands used throughout the requirements and assertion tables. The parameter X may represent any combination of operands, terms, and operators.

Table 5.3 - Assertion Syntax: Operand Definitions

Operand Definitions		
Operand	**Name**	**Description**
All(X)	All Occurrences	Returns all occurrences of X.

ASCII(X)	ASCII Values	Specifies that all values represented by X are ASCII values. Ex. ASCII(a) is 0x61
Bytes(X)	Byte Data	Returns the set of bytes contained in X.
Count(X)	Count Occurrences	Returns the number of occurrences of X.
DataLength(X)	Length Of (without Special Characters)	Returns the length of X without counting the characters ASCII(US, RS, FS).
FieldNumber(X)	Field Number	Returns the field number of X.
First(X)	First Occurrence	Returns the first occurrence of X.
For(X EQ A to B) {Expression(s)}	For Loop	Evaluates each Expression for the range specified by A to B.
ForEach(X) {Expression(s)}	For Each	Evaluates each Expression for every occurrence of X found.
Last(X)	Last Occurrence	Returns the last occurrence of X.
Length(X)	Length Of	Returns the length of X.
Max(X)	Maximum Value	Returns the maximum value in the set X.
Min(X)	Minimum Value	Returns the minimum value in the set X.
Next(X)	Next Occurrence	Returns the next occurrence of X. Only for use within ForEach Operand's Expression(s).
Pair(A,B) of X	Pair	Returns all pairs of X. Only for use as a parameter in a ForEach Operand.
ParentField(X)	Parent Field	Returns the Field that contains X.
ParentRecord(X)	Parent Record	Returns the Record that contains X.
Present(X)	Value Present	Returns TRUE if X is present, FALSE otherwise. For subfields in Traditional Encoding, the US and RS separators are always present. Therefore the Present(X) operand returns TRUE if the value between the separators is present.
Previous(X)	Previous Occurrence	Returns the previous occurrence of X. Only for use within ForEach Operand's Expression(s).
Second(X)	Second Occurrence	Returns the second occurrence of X.
Type(X)	Record Type	Returns the Record Type of X.
Var(X) {Selection Statement}	Variable	Assigns the entity specified by the Selection Statement to the name X. The assignment is valid for the remainder of the assertion text.
XElm(X)	XML Element	Returns the XML Element with name X.

5.6   Tables of requirements and assertions - Table headers

The following describe the headings of the tables of requirements and assertions found in Section 6:

- **Requirement ID**: Defines a unique identifier for the requirement and associated assertion or set of assertions. It provides reference to the type of requirement (e.g., transaction, record, and field). The Requirement ID is in the form of "Type: Description" where type may be "Transaction", "Record", or "Field". For requirements found in Annex B of the AN-2011 standard, the Requirement ID is preceded by "Traditional-".

- **Reference in Base Standard**: Identifies the clause (or section) where the requirement is included in the AN-2011 standard. In some cases the reference includes additional information such as a Table number.

- **Requirement Summary**: Provides a summary of the requirement detailed as textual information or an interpretation of the requirement in the standard. It provides the essentials of the requirement but may not provide all the text necessary to understand it. The < > operator is used in the Requirement Summary column of the tables to represent text not found in the standard, but that may help indicate what requirement is being represented.

- **Level**: Indicates whether Level 1 or Level 2 conformance testing is required to address the assertion identified in the Assertion ID column of the same row. Level 3 conformance tests are indicated only when necessary to show that the requirement is not currently testable or addressed.

- **Status**: Reflects the status specified in the AN-2011 standard:
  - M: Mandatory
  - O: Optional
  - D: Dependent
  - M⇑: Mandatory within the optional field/subfield
  - O⇑: Optional within the optional field/subfield
  - -: Varying statuses. The assertion addresses many fields or subfields of multiple statuses.

- **Assertion ID**: Defines an identifier of a specific test assertion within the set of test assertions associated with a requirement.

- **Test Assertion**: Provides, whenever possible, a mathematical equation or a procedure using the language specified by the operators, operands, and terms.
  - The < > operator is used to contain plain text whenever a mathematical formula or simple procedure cannot be detailed.

- **Test Note**: Contains the ID of the test note. Test notes provide additional information related to the assertion and are included below the tables.

- **Implementation Support**: Denotes a supplier's implementation support of a particular requirement ("Y"/"N"). A note can follow the table when providing more details of implementation support (or the lack of it) is required.

- **Supported Range**: Indicates a range of values supported, especially when it is different than the full range of values specified in the standard. When an information item is specified as a single value, or does not address a range of values, a N/A should be used.

- **Test Result**: This column is used to denote the test results. For file and record-level results the results are either "Pass" or "Fail". The field-level results should be indicated as "Ok", "Error", "Warning" and "Note". Explanatory notes can be added below the table.

- **Applicability**: This table header indicates which assertions differ (in values required or conditions) between Traditional and NIEM encoding. This table header does not indicate which assertions are addressed by the XML Schema and which will need to be addressed in code. Valid values are:

  - T: The assertion only applies to the Traditional encoding as described in Annex B.

  - X: The assertion only applies to the NIEM-conformant (XML) encoding as described in Annex C.

  - B: The assertion is applicable to both Traditional and NIEM (XML) encoding.

    - Following the conventions in the AN-2011 standard, test Assertions are expressed using constructs (fields, records, etc.) found in Traditional encoding (such as xx.002 for the second field of each record type). The same assertion applies for the XML elements that correspond to the Traditional constructs. For example, 10.006 in Traditional Encoding corresponds to XML Element <biom:ImageHorizontalLineLengthPixelQuantity>.

  - Some assertions reference subfields, however, NIEM encoding uses nested elements. Expression of Test Assertions that include subfields in the XML encoding requires further review. These assertions are listed with the following applicability values:

    - X* indicates that the assertion applies only to NIEM-conformant (XML) encoding.

    - B* indicates that the assertion is applicable to both Traditional and NIEM (XML) encodings.

6 Tables of requirements and assertions for AN-2011

6.1 Scope of requirements and test assertions specified in the standard

The CTM specifies requirements and test assertions for the following Sections and selected Record Types of the ANSI/NIST-ITL 2011 standard:

- Section 5: Data Conventions
- Section 7: Information Associated with Several Record Types
- Section 8.1 Record Type-1: Transaction information record
- Section 8.4 Record Type-4: Grayscale fingerprint image
- Section 8.10Record Type-10: Facial, other body part and SMT image record
- Section 8.13 Record Type-13: Friction-ridge latent image record
- Section 8.14 Record Type-14: Fingerprint image record
- Section 8.15 Record Type-15: Palm print image record
- Section 8.17 Record Type-17: Iris image record
- Annex B: Traditional Encoding

Tables for the deprecated Record Types 3, 5, and 6 are also included to check for nonexistence of these Record Types. Additionally an assertion is specified that checks for the nonexistence of reserved Record Types 11, 12, and 22 through 97.

6.2 Field Definitions and Structures

ANSI/NIST-ITL 1-2011 contains several field types:

- Single Information Item (Field with data)
- Multiple Information Items (Field with multiple Information Items)
- Subfields Repeating Sets of Info Items
- Subfields Repeating Values

The tables of requirements and assertions represent all field types as a field that contains a list of one or more subfields, each of which contains a list of one or more information items. Fig. 6.1 is a representation of how each field type is represented in the tables.

11

- Single Information Item: Field with one subfield containing one information item.
- Multiple Information Items: Field with one subfield containing multiple information items.
- Subfields Repeating Sets of Information Items: Field with one or more subfields, each containing sets of one or more information items.
- Subfields Repeating Values: Field with one or more subfields, each containing one information item.

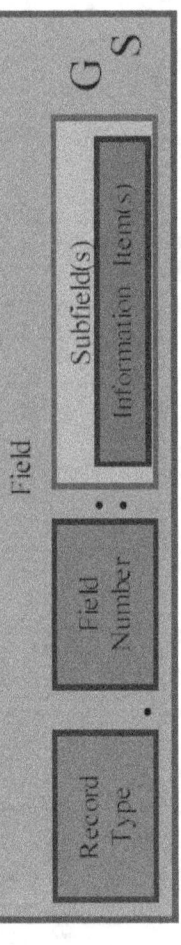

Figure 6.1 - Generic AN-2011 Field Structure

Table 6.1 - Assertions for Data Conventions

Requirement ID	Reference in Base Standard	Requirement Summary	Level	Status	Assertion ID	Test Assertion	Test Note	Implementation Support	Supported Range	Test Result	Applicability
						5: Data Conventions					
Transaction: Required Record Types	5.1, Table 3	There shall be at least one other record type from Table 3 accompanying a Record Type-1.	1	M	Transaction-Required Records	Present(Record ST Type(Record) EQ 1) AND Present(Record ST Type(Record) MO [2 to 99]) AND NOT MO [3,5,6,11,12,22 to 97])					T
			1	M	NIEM-Transaction-Required Records	Count(XElm(itl:PackageInformationRecord) in Transaction) EQ 1 AND Count(Records) GTE 2					X
Transaction: Single	5.1	All records in a transaction shall pertain to a single subject. Biometric data used to	3	M	Transaction-Single	<Unsupported>	t-1				B

12

Subject					Subject			
		identify another individual requires a separate transaction.						
Transaction: Records Transmitted Together	5.1	All of the records belonging to a single transaction shall be transmitted together.	3	M	Transaction-Records Together	<Unsupported>	t-1	B
Transaction: Record Occurrences	5.1	There may be multiple records in a transaction of each record type other than Type-1.	1	M	Transaction-Record Occurrences	TRUE		B
Transaction: Size	5.2	Although the 2007 and 2008 versions of the standard stated "… there is no upper limit on the number of logical records that may be present in a file…" there was an effective upper limit due to the field size limits specified in the 2007 version (but not the 2008 version). This limit was 3 ASCII2 characters for the information item holding the total number of records of type 2 through 99; thus an upper limit of 999 such records. With the addition of a Type-1 record, the maximum number of records in a transaction was thus restricted to 1000. This upper limit of 1000 records is maintained in this version of the standard to ensure backward compatibility with the 2007 version.	1	M	Transaction-Size	Count(Records in Transaction) LTE 1000		B
Transaction: Reserved Records	5.3 Table 3	11 Voice Data (future) 12 Dental Record (future) 22-97 reserved for future use.	1	M	Records-Reserved	NOT Present(Records ST Type(Records) MO [11,12,22 to 97])		T
			1	M	NIEM-Records Reserved	{All(XElm(<biom:RecordCategoryCode>)]} NOT MO [11,12,22 to 97]		X
Transaction: Type1-Occurrences	5.3.1	Transmissions to be exchanged are required to contain one and only one Type-1 record per transaction.	1	M	Type1-Occur Once	Count(Records in Transaction ST Type(Record) EQ 1) EQ 1		T
			1	M	NIEM-Type1-Occur Once	Count(XElm(iti:PackageInformationRecord)) EQ 1		X
Transaction: Type1-Record_First	5.3.1	The Type-1 record shall always be the first record within the transaction.	1	M	Type1-First	Type(First(Record in Transaction)) EQ 1		T
			1	M	NIEM-Type1-First	First(Record in Transaction) EQ XElm(iti:PackageInformationRecord)		X
Transaction: Type1- One More Record	5.3.1	At least one more record shall be present in the file.	1	M	Type1-One More Record	<See Requirement ID "Transaction: Required Record Types">	t-2	T
Record: Type1-	5.3.1	The Type-1 record shall provide information describing type and use or		M	Type1-Contents	<The test assertions are included under field testing for Record Type-1: Transaction	t-2	

Category	Section		M	Type	Assertion / Description	Code		
Contents					<u>Information Record ></u>			
Record: Type2-Contents	5.3.2		M	Type2-Contents	purpose for the transaction involved, a listing of each record included in the transaction, the originator or source of the physical record, and other useful and required information items. Type-2 records shall contain user-defined textual fields providing identification and descriptive information associated with the subject of the transaction.	< The test assertions for this type may not be supported in this version of the CTM. If they are supported, they are included under field testing for Record Type-2: User-defined descriptive text record >	t-2	
Record: Type2-DOM/APS	5.3.2	3	M	Type2-DOM/APS	Each entry in a Type-2 record shall have a definition and format that is listed with the Domain owner. Data contained in this record shall conform in format and content to the specifications of the domain name(s) as listed in Field 1.013 Domain name / DOM found in the Type-1 record, if that field is in the transaction. The default domain is NORAM. Field 1.016 Application profile specifications / APS allows the user to indicate conformance to multiple specifications. If Field 1.016 is specified, the Type-2 record must conform to each of the application profiles.	<Unsupported>	t-3	B
Transaction: Type3-Deprecated	5.3.3, Table 3		M	Type3-Unsupported	Record Type-3 shall not be contained in transactions conforming to this version of the standard.	<The test assertions are included under field testing for <u>Record Type-3</u>: DEPRECATED.>	t-2	
Record: Type4-Contents	5.3.4		M	Type4-Contents	Type-4 records were designed to convey fingerprint images captured by an Automated Fingerprint Identification System (AFIS) live-scan reader, or other image capture devices operating at a nominal scanning resolution of 500 pixels per inch (ppi). Many systems still use this record type and it will remain an integral part of the standard.	<The test assertions are included under field testing for <u>Record Type-4</u>: <u>Grayscale fingerprint image</u>.>	t-2	
Transaction: Type5-Deprecated	5.3.5, Table 3		M	Type5-Unsupported	Record Type-5 shall not be contained in transactions conforming to this version of the standard.	<The test assertions are included under field testing for <u>Record Type-5</u>: DEPRECATED.>	t-2	
Transaction: Type6-Deprecated	5.3.6, Table 3		M	Type6-Unsupported	Record Type-6 shall not be contained in transactions conforming to this version of the standard.	<The test assertions are included under field testing for <u>Record Type-6</u>: DEPRECATED.>	t-2	
Transaction: Type7-	5.3.7		M	Type7-Contents	Type-7 is a legacy record type. It was intended as a temporary measure to	<The test assertions for this type may not be supported in this version of the CTM. If they are	t-2	

Contents		enable the exchange of image data that would be defined by specific record types in later versions of the standard. Since some older systems still use this record type, it is included in the standard.			supported, they are included under field testing for Record Type-7: User-defined image record .>	
Record: Type8-Contents	5.3.8	Type-8 records shall be used for scanned binary or vectored signature image data. Each Type-8 record shall contain data representing the signature of the subject from whom the biometric sample is being collected and/or the operator capturing biometric data.	M	Type8-Contents	<The test assertions for this type may not be supported in this version of the CTM. If they are supported, they are included under field testing for Record Type-8: Signature image record.>	t-2
Record: Type9-Contents	5.3.9	Type-9 records shall contain and be used to exchange minutiae or other friction ridge feature data. Each record shall represent the processed (automated and/or manual) image data from which the characteristics are stated. The primary use of this record type shall be for remote searching of latent prints. New to this version of the standard is the Extended Feature Set (EFS) for latent print markups. There is also a capability to have additional vendor-specified feature sets. Workstation logs may also now be transmitted in this record type.	M	Type9-Contents	<The test assertions for this type may not be supported in this version of the CTM. If they are supported, they are included under field testing for Record Type-9: Minutiae data record .>	t-2
Record: Type10-Contents	5.3.10, Table 58	Type-10 image records shall contain and be used to exchange textual and image data from the face, scars, (needle) marks, and tattoos (SMT). New to this version of the standard is the extension of the record type to handle images of other body parts. See Table 58 for a list of the images types possible in a Type-10 record. Textual and analytic information pertinent to the digitized image is also contained in this record type.	M	Type10-Contents	<The test assertions are included under field testing for Record Type-10: Facial other body part and SMT image record.>	t-2
Transaction: Type11-Reserved	5.3.11	Type-11 records are reserved for future use.	M	NA	<See Requirement ID "Transaction: Reserved Records">	t-2
Transaction: Type12-Reserved	5.3.12	Type-12 records are reserved for future use.	M	NA	<See Requirement ID "Transaction: Reserved Records">	t-2
Record: Type13-	5.3.13	Type-13 image records shall contain and be used to exchange variable-resolution	M	Type13-Contents	<The test assertions are included under field testing for Record Type-13: Friction-ridge latent	t-2

Contents		latent friction ridge image data (fingerprint, palmprint and/or plantar) together with fixed and user defined textual information fields pertinent to the digitized image. In all cases, the scanning resolution for latent images shall be at least 39.37 ppmm (1000 ppi). The variable resolution latent image data contained in the Type-13 record shall be uncompressed or may be the output from a lossless compression algorithm.			_image record_ . >	
Record: Type14-Contents	5.3.14	Type-14 image records shall contain fingerprint image data. It should be noted that as the class resolution is increased, more detailed ridge and structure information becomes available in the fingerprint image. However, in all cases the class resolution shall be at least 19.69 ppmm (500 ppi).	M	Type14-Contents	<The test assertions are included under field testing for Record Type-14: Fingerprint image record.>	t-2
Record: Type15-Contents	5.3.15	Type-15 image records shall contain and be used to exchange palm print image data together with fixed and user-defined textual information fields pertinent to the digitized image. ...in all cases the class resolution shall be at least 19.69 ppmm (500 ppi) The variable-resolution palm print image data contained in the Type-15 record may be in a compressed form.	M	Type15-Contents	<The test assertions for this type are included under field testing for Record Type-15: Palm print image record >	t-2
Record: Type16-Contents	5.3.16	The Type-16 image record is designed for developmental purposes and for the exchange of miscellaneous images. This record shall contain and be used to exchange image data together with textual information fields pertinent to the digitized image.	M	Type16-Contents	<The test assertions for this type may not be supported in this version of the CTM. If they are supported, they are included under field testing for Record Type-16: User-defined testing image record.>	t-2
Record: Type17-Contents	5.3.17	Type-17 image records shall contain iris image data. Field 17.018 (Global unique identifier) from the 2007 and 2008 version of the standard has been deprecated in this version.	M	Type17-Contents	<The test assertions are included under field testing for Record Type-17: Iris image record .>	t-2
Record: Type18-Contents	5.3.18	The Type-18 record (new to this version of the standard) shall contain and be used to exchange DNA and related data. It was developed to provide a basic level of interoperability with the draft format of the _ISO/IEC 19794-14 DNA data_	M	Type18-Contents	<The test assertions for this type may not be supported in this version of the CTM. If they are supported, they are included under field testing for Record Type-18: DNA record .>	t-2

		interchange format. With full consideration to privacy, this standard only uses the non-coding regions of DNA. The regions of the DNA that encode phenotypic information are deliberately avoided.				
Record: Type19-Contents	5.3.19	Type-19 image records (new to this version of the standard) shall contain and be used to exchange variable-resolution plantar print image data together with fixed and user-defined textual information fields pertinent to the digitized image. ... in all cases the scanning resolution used to capture a plantar image shall be at least as great as the minimum scanning resolution of 19 69 ppmm (500 ppi). The variable-resolution plantar image data contained in the Type-19 record may be in a compressed form.	M	Type19-Contents	\<The test assertions for this type may not be supported in this version of the CTM. If they are supported, they are included under field testing for Record Type-19: Plantar record.\>	t-2
Record: Type20-Contents	5.3.20	The Type-20 record (new to this version of the standard) shall contain the source representation(s) from which other Record Types were derived. Typically, one Type-20 source representation is used to generate one or more representations for use in other record types. When a source representation (in a Type-20 record) is processed and the derived representation is to be used as the source for further derivations, then the derived representation is contained in a Type-20 record.	M	Type20-Contents	\<The test assertions for this type may not be supported in this version of the CTM. If they are supported, they are included under field testing for Record Type-20: Source representation record.\>	t-2
Record: Type21-Contents	5.3.21	The Type-21 record shall contain an associated context image, audio / visual recording or other related data. This record type does NOT contain information used to derive biometric information contained in other records. Record Type-20 serves that function. Record Type-21 may be used to convey contextual information, such as an image of the area where latent fingerprints were captured.	M	Type21-Contents	\<The test assertions for this type may not be supported in this version of the CTM. If they are supported, they are included under field testing for Record Type-21: Associated context record.\>	t-2
Record: Type98-Contents	5.3.22	The Type-98 record shall contain security information that allows for the assurance of the authenticity and/or integrity of the transaction, including such information as	M	Type98-Contents	\<The test assertions for this type may not be supported in this version of the CTM. If they are supported, they are included under field testing for Record Type-98: Information assurance	t-2

Name	Section	Description	#	Req	Short Name	Requirement / Test Assertion	Code
		binary data hashes, attributes for audit or identification purposes, and digital signatures.				record.>	
Record: Type99-Contents	5.3.23	Type-99 records shall contain and be used to exchange biometric data that is not supported by other ANSI/NIST-ITL record types. This provides a basic level of interoperability and harmonization with other biometric interchange formats. This is accomplished by using a basic record structure that is conformant with *ANSI INCITS 398-2005, the Common Biometric Exchange Formats Framework (CBEFF)* and a biometric data block specification registered with the International Biometrics Industry Association (IBIA).		M	Type99-Contents	<The test assertions for this type may not be supported in this version of the CTM. If they are supported, they are included under field testing for Record Type-99: CBEFF biometric data record.>	t-2
Transaction: Deprecated Entities	5.4	Deprecated records for this version are Record Types 3, 5 and 6. Field 17.018 is deprecated. There are two deprecated values in Field 17.016: Image property code / IPC (2: for interlace frame, and 3: for interlace field).	1	M	Deprecated Entities	<See Requirement IDs: "Field: Type 17-CondCode" and "Field: 17.016-Image Property Code Value"> <See Sections "8.3: Record Type-3: DEPRECATED", "8.5: Record Type-5: DEPRECATED", and "8.6: Record Type-6: DEPRECATED".>	t-2
Transaction: Legacy Entities	5.4	There is a special category called 'legacy' for a record type, field, subfield, information item or value that was valid in previous versions of the standard, but shall not be used for new data. 'Legacy' indicates that if there is existing data using this record type, field, information item or value it may still be transmitted in a transaction conformant to this version of the standard. In this version 'legacy' applies to Fields 9.005 through 9.012, Field 10.022 and to the value '1' in Table 4 Character encoding.	3	M	Legacy Entities	<Warning provided only. It is not feasible to determine if the data is new or legacy. However, a warning will be displayed if legacy entities are present for the supported record types. >	B
Transaction: Reserved Character Types	5.6, Table 93	<Table 93 specifies a set of reserved character types.>	1	M	Reserved Character Types	ForEach(Record in Transaction ST Type(Record) NOT MO [4, 7, 8]) { ForEach(Field in Record ST FieldNumber(Field) NEQ 999) { ForEach(Subfield in Field) { Bytes(All(InfoItems in Subfield)) NOT MO [0x02, 0x03, 0x1C, 0x1D, 0x1E, 0x1F]	B

18

Table 6.2 - Assertions for Implementation Domain & Application Profiles

Requireme nt ID	Referen ce in Base Standar d	Requirement Summary	Lev el	Stat u I	Assertion ID	Test Assertion	Test Note	Implemen tation Support	Supporte d Range	Test Result	Applicab ility
Record: Type1-ASCII	5.6, Table 93	Record Type-1 shall always be recorded in all encodings using the characters that can be represented by the 7-bit American National Standard Code for Information Interchange (ASCII) found in table 93 with the exception of the reserved values.	1	M	Type1-ASCII	ForEach(Field in Record ST Type(Record) EQ 1) { {Bytes(Field)} MO [0x20 to 0x7E] }					T
	C.4.1	For compatibility with existing implementations of the standard, implementers may wish to limit content to the 128 characters that can be represented by 7-bit ASCII. Nevertheless, senders and receivers of XML packages using this standard may agree on other character sets, including international character sets.	1	M	NIEM-Type1-User Defined	TRUE	t-4				X
Transaction: Encoding-Base64	5.6	Base-64 shall be used for converting non-ASCII text into ASCII form, where required and noted in the standard.	1	M	Data-Encoding-Base-64	<Unsupported>	t-4				B
Field: Encoding-CharSets	5.6, Table 4	Field 1.015 Character encoding/DCS is an optional field that allows the user to specify an alternate character encoding… Field 1.015 Character encoding/DCS contains three information items: the character encoding set index/CSI, the character encoding set name/CSN, and the character encoding set version/CSV. The first two items (CSI and CSN) are selected from the appropriate columns of Table 4 .	3	O	Data-Encoding-CharSets	<Unsupported>	t-4				B

19

6: Implementation Domain and Application Profiles

	s					
Field: Domain	6	3	O	Fields-Domain	<Unsupported>	t-3	B
				An implementation domain, coded in Field 1.013 Domain name / DOM of a Type-1 record as an optional field, is a group of agencies or organizations that have agreed to use preassigned data fields with specific meanings (typically in Record Type-2) for exchanging information unique to their installations. The implementation domain is usually understood to be the primary application profile of the standard.			
Field: APS	6	3	O	Fields-APS	<Unsupported>	t-3	B
				Field 1.016 Application profile specifications / APS allows multiple application profiles to be referenced. The organization responsible for the profile, the profile name and its version are all mandatory for each application profile specified. A transaction must conform to each profile that is included in this field.			

20

Table 6.3 - Assertions for Information Associated with Several Records

Requirement tID	Reference in Base Standard	Requirement Summary	Level I	Status	Assertion ID	Test Assertion	Test Note	Implementation Support	Supported Range	Test Result	Applicability
						7: Information Associated with Several Records					
Field: xx.001-Record Header	7.1	The record header appears as the first field (xx.001) in each Record Type. It contains information particular to the encoding format chosen, in order to enable proper reading of the record. In Traditional encoding, this field contains the record length in bytes (including all information separators).	1	M	xx.001-First	ForEach(Record in Transaction) { FieldNumber(First(Field in Record)) EQ 1 }					T
			1	M	NIEM-RecordCategory First	ForEach(Record in Transaction) First(Field in Record) XElm(<biom:RecordCategoryCode>) }					X
			2	M	xx.001-Value	ForEach(Record in Transaction) {Record.001} EQ Length (Record) }					T
	C.4.7, 7.1	Record Length. There is no corresponding XML element. See Section 7.1. In NIEM-conformant XML encoding, this field contains the *RecordCategoryCode*, which is the numeric representation of the Record Type.	1	M	NIEM-xx.001-Record Category	< These assertions are included under field testing for the associated record types >	t-2				
Transaction: Record Length	7.1	In the 2007 version of the standard, the record length was unrestricted for Record Type-1. It was a maximum value having up to 4-bytes in ASCII representation for Record Types 4 and 7 and 8.	1	M	Record Lengths	< These assertions are included under field testing for the associated record types >	t-2				

Field: xx.999-Reserved		M	T	t-2	X
7.2	For Record Types 9 and above it was restricted to 8 characters (99,999,999). These values are retained in this version for Traditional encoding.				
1	Field xx.999 is reserved in Record Types 10 and above for data associated with the record that is described in the other fields of the record. It is mandatory in most of these record types (It does not appear in Type-18 or Type-98). Only in Record Types 14, 15, 17 and 19 is it possible for Field xx.999 to be optional – when an amputated or missing body part is noted in the appropriate field in those record types.	M xx.999-Reserved	ForEach(Field ST FieldNumber(Field) EQ 999)	< These assertions are included under field testing for the associated record types >	
		M xx.999-RecordTypes	ForEach(Field ST FieldNumber(Field) EQ 999) { Type(ParentRecord(Field)) GTE 10 AND NOT MO [18, 98] }		
1		M NIEM-ImageRecordCategoryCode	ForEach(Field ST Field EQ XElm(biom:FaceImage) OR XElm(biom:PhysicalFeatureImage)) {XElm(biom:RecordCategoryCode) in ParentRecord(Field)} EQ 10}

AND

ForEach(Field ST Field EQ XElm(biom:PlantarImage) OR XElm(biom:FrictionRidgeImage)) {XElm(biom:RecordCategoryCode) in ParentRecord(Field)} EQ 13}

AND

ForEach(Field ST Field EQ XElm(biom:FingerImpressionImage) {XElm(biom:RecordCategoryCode) in ParentRecord(Field)} EQ 13 OR 14}

AND

ForEach(Field ST Field EQ XElm(biom:PalmprintImage) {XElm(biom:RecordCategoryCode) in ParentRecord(Field)} EQ 13 OR 15}

AND

ForEach(Field ST Field EQ | | |

			Num	M	Name	Logic	Type
Field: xx.002-IDC	7.3.1	Each of the records present in a transaction, with the exception of the Type-1 record, shall include a field (xx.002) containing the information designation character / IDC6. The value of the IDC shall be a sequentially assigned positive integer starting from zero and incremented by one up to a maximum of 99. IDC references are stated in Type-1 Field 1.003 Transaction content / CNT and shall be used to relate information items in the CNT field of the Type-1 record to the other records in the transaction Two or more records may share a single IDC solely to identify and link together records that pertain to different representations of the same biometric trait.				XElm(biom:IrisImage) { {XElm(biom:RecordCategoryCode) in ParentRecord(Field)} EQ 17}	T
			1	M	xx.002-Exists	ForEach(Record ST Type(Record) NEQ 1) { Present(Record.002) }	X
			1	M	NIEM-IDCExists	ForEach(Record ST Record NEQ XElm(itl:PackageInformationRecord)) { Present(XElm(biom:ImageReferenceIdentification)) }	T
			1	M	xx.002-Second	ForEach(Record ST Type(Record) NEQ 1) FieldNumber(Second(Field in Record)) EQ 2 }	X
			1	M	NIEM-IDC-Second	ForEach(Record ST Record NEQ XElm(itl:PackageInformationRecord)) { Second(Field in Record) EQ XElm(biom:ImageReferenceIdentification) }	T
			2	M	xx.002-IDCSeqValues	Var(IDC_Fields) All(Fields in Records ST Type(Records) NEQ 1 AND FieldNumber(Fields) EQ 2) } {First(First in IDC_Fields)} EQ 0 AND ForEach(Field in IDC_Fields) {{Next(Field)} LTE <Current Maximum IDC Value> +1 AND {Field} MO [Integers] }	
			1	M	xx.002-Value	ForEach(Record ST Type(Record) NEQ 4 OR 8) {	B

Field	Section	Description	#	M/O	Field ID	Assertion / Definition	Test Note	Status
						{Record.002} MO [0 to 99] }		X
			1	M	NIEM-IDCSeqValues	Var(IDC_Fields) All({XElm(biom:ImageReferenceIdentification)}) {First(Field in IDC_Fields)} EQ 0 AND ForEach(Field in IDC_Fields) { {Next(Field)} LTE <Current Maximum IDC Value> +1 }		
			2	M	xx.002-IDCRelate 1.003-CNT	<See tables in test note>	t-5	B
						<These assertions are included under field testing for Record Type-1: Transaction Information Record >	t-2	B
Field: xx.002-IDCImages	7.3.1	Two or more image records may share a single IDC only when they are enhancements of a single image; such transformations shall have identical dimensions.	2	M	xx.002-SameDimension	ForEach(Pair(A,B) of Records <with matching IDC fields>) { {A.006} EQ {B.006} AND {A.007} EQ {B.007} }		B
			3	M	xx.002-SameImage	<Unsupported: Not feasible to test if the samples are from the same image, only that the samples come from the same type of biometric trait (see Field: xx.002-IDC)>	t-1	
Field: xx.997-SOR	7.3, 7.3.2	...optional field xx.997 is allowed in biometric data sample Record Types 10 and above that could have the biometric sample derived from a source representation in Record Type-20. The biometric data is stored in Field xx.999. Record Type-18 (DNA) does not contain a	1	O	xx.997-RecordTypes	ForEach(Field ST Type(Field) EQ 997) { Type(ParentRecord(Field)) GTE 10 AND NOT MO [18,20, 21,98] }		T
			1	O	NIEM-SOR-	ForEach(Field ST Field EQ		X

24

Field	Section	Description	#	M/O	Name	Condition	
		field xx.997, since it does not contain a field 18.999. Record Type-98 does not contain this field, since that is not a biometric data record type. Record Type-21 does not contain biometric data and thus does not include field xx.997. This field is comprised of one mandatory and one optional information item, as described below.			RecordTypes	XElm(biom:SourceRepresentation)) { {XElm(biom:RecordCategoryCode) in ParentRecord(Field)} GTE 10 AND NOT MO [18, 20, 21, 98] }	B*
			1	O	xx.997-SubfieldCount	ForEach(Field ST FieldNumber(Field) EQ 997) { Count(Subfields in Field) MO [1 to 255] }	
Field: xx.997-SOR-SRN	7.3, 7.3.2.1	The first information item contains the source representation number / SRN. This is mandatory for each Field xx 997. It contains an index to a particular instance of a Type-20 record in the transaction. This same index value appears in the appropriate instance of Record Type-20 as Field 20.021: Source representation number / SRN. The value of the SRN shall be a sequentially assigned positive integer starting from one and incremented by one, not to exceed 255.	2	M ⇑	xx.997-SOR-SRN	ForEach(Field ST FieldNumber(Field) EQ 997) { ForEach(Subfield in Field) { Present(Record in Transaction ST Type(Record) EQ 20 AND {Record.021} EQ {InfoItem:1 in Subfield}) }	B*
			2	O	xx.997-SOR-SRN-SeqValues	Var(SOR_Fields) All(Fields ST FieldNumber(Fields) EQ 997) { Var(MaxSOR)[2] {InfoItem:1 in Subfield:1 in First(Field in SOR_Fields)} EQ 1 AND ForEach(Field in SOR_Fields) { ForEach(Subfield in Field) { {InfoItem:1 in Subfield} LTE {MaxSOR} AND IF {InfoItem:1 in Subfield} EQ {MaxSOR} THEN Var(MaxSOR) {MaxSOR + 1}	B*
			1	O	xx.997-	ForEach(Field ST FieldNumber(Field) EQ 997)	B*

Field	Ref	Description	#	O/M	Subfield	Definition		Type
					SRN-OneTo255	{ ForEach(Subfield in Field) { [InfoItem:1 in Subfield} MO [1 to 255] } }		B*
Field: xx.997-SOR-RSP	7.3, 7.3.2.2	The second information item in Field xx.997 is optional. It is the reference segment position / RSP. It contains the index to a particular set of segmentation coordinates of the source representation (There may be more than one segment, such as from an audio / visual recording, with different frames yielding input for separate biometric data record instances in the same transaction). This same segmentation index value appears in Record Type-20 as the reference segment position / RSP in Field 20.016: Segments / SEG. There may be up to 99 segments listed in Field 20.016, but only the segment used to produce the biometric data contained in Field xx.999 of the particular instance of Record Type-xx is identified in Field xx.997.	2	O ⇑	xx.997-SOR-RSP	ForEach(Field ST FieldNumber(Field) EQ 997) { ForEach(Subfield in Field) { IF(Present(InfoItem 2 in Subfield) { Present(Record in Transaction ST Type(Record) EQ 20 AND {Record.021} EQ {InfoItem:1 in Subfield} AND Present Subfield in{Record.016} ST InfoItem:1 in Subfield EQ [InfoItem:2 in Field) } }}		B*
			3	O	xx.997-SOR-RSP-CorrectSegment	<Unsupported: Not feasible to test if the RSP that matches was used to produce the biometric data. >	t-1	B*
			1	O	xx.997-RSP-OneTo99	ForEach(Field ST FieldNumber(Field) EQ 997) { ForEach(Subfield in Field) { [InfoItem:2 in Subfield] MO [1 to 99] } }		B*
			1	O	20.016-SubfieldCount	Count(Subfields in 20.016) MO [1 to 99] AND MO [Integers]		B*
Field: xx.995-ASC	7.3, 7.3.3	New to this version of the standard, optional field xx.995 is contained in biometric data sample Record Types 10 and above that may have instances of Record Type-21 linked to it...	1	O	xx.995-RecordTypes	ForEach(Field ST FieldNumber(Field) EQ 995) { Type(ParentRecord(Field)) GTE 10 AND NOT MO [21, 98] }		T
		This field consists of a maximum of 255 repeating subfields, each of which contains two information items, as described below.	1	O	NIEM-ASC-RecordTypes	ForEach(Field ST Field EQ XElm(biom:AssociatedContext) {XElm(biom:RecordCategoryCode) in ParentRecord(Field)} GTE 10 AND NOT MO [21,		X

Ref	Field	#	M/O	Subfield	Rule	B*
					98]	B*
		1	O	xx.995-SubfieldCount	ForEach(Field ST FieldNumber(Field) EQ 995) { Count(Subfields in Field) MO [1 to 255] }	B*
7.3, 7.3.3.1	Field: xx.995-ASC-ACN	2	M ⇑	xx.995-ASC-ACN	ForEach(Field ST FieldNumber(Field) EQ 995) { ForEach(Subfield in Field) { Present(Record in Transaction ST Type(Record) EQ 21 AND {Record.021} EQ { InfoItem:1 in Subfield}) } }	
		2	O	xx.995-ASC-ACN-SeqValues	Var(ASC_Fields) All(Fields in Transaction ST FieldNumber(Fields) EQ 995) } Var(MaxASC){2} {InfoItem:1 in Subfield:1 in First(Field in ASC_Fields)) EQ 1 AND ForEach(Field in ASC_Fields) { ForEach(Subfield in Field) { IF({InfoItem:1 in Subfield} EQ {MaxASC} THEN Var(MaxASC) {MaxASC + 1} } }	B*
		1	O	xx.995-ACN-Value	ForEach(Field ST FieldNumber(Field) EQ 995) { ForEach(Subfield in Field) { InfoItem:1 in Subfield} MO [1 to 255] }	B*

Description (Field: xx.995-ASC-ACN, 7.3, 7.3.3.1):

The first information item contains the associated context number / ACN for a particular Record Type-21. This is mandatory for each Field xx.995, when the field is used. It contains an index to a particular instance of a Type-21 record in the transaction. This same index value appears in the appropriate instance of Record Type-21 as Field 21.021: Associated context number / ACN. The value of the ACN shall be a sequentially assigned a positive integer starting from one and incremented by one, not to exceed 255.

Field	Section	Description	#	Code	Identifier	Logic	t-1	Level

Field: xx.995-ASC-ASP	7.3, 7.3.3.2	The second information item in Field xx.995 is optional. It is the associated segment position / ASP. It contains the index to a particular set of segmentation coordinates of the associated context data. This same segmentation index value appears in Record Type-21 as the associated segment position / ASP in Field 21.016: Segments / SEG. There may be up to 99 segments listed in Field 21.016, but only the relevant segment is contained in Field xx.995.	2	O ⇐	xx.995-ASC-ASP	ForEach(Field ST FieldNumber(Field) EQ 995)
						{
						ForEach(Subfield in Field)
						{
						IF(Present(InfoItem 2 in Subfield)
						{
						Present(Record in Transaction ST Type(Record) EQ 21
						AND
						{Record.021} EQ {InfoItem:1 in Subfield}
						AND
						Present Subfield in {Record.016} ST InfoItem:1 in Subfield EQ {InfoItem:2 in Subfield)
						}
						}
						}
								B*

			1	O	xx.995-ASP-Value	ForEach(Field ST FieldNumber(Field) EQ 995)
						{
						ForEach(Subfield in Field)
						{
						{InfoItem:2 in Subfield} MO [1 to 99]
						}
						}
								B*

			3	O ⇐	xx.995-ASC-ASP-CorrectSegment	<Unsupported: Not feasible to test if the ASP that matches is the one that is relevant >	t-1	B

Field: 10.039-RefNum	7.3, 7.3.4	There may be several Type-10 images of a particular part of the body. For instance, a photograph of a tattoo may cover the entire tattoo. Another may be a zoom-in shot of a portion of the tattoo. In order to link these two images, the same index number is assigned to Field 10.039: Type-10 reference number / T10, which is new to this version of the standard. Note that these images would have different IDC values.	2	D	10.039-T10-DiffIDC	IF Count(Records ST Type(Records) EQ 10) GT 1
						THEN
						ForEach(Pair (A,B) of Records ST Type(Records) EQ 10)
						{
						IF {A.039} EQ {B.039}
						THEN
						{A.002} NEQ {B.002}
						}
								T

			2	D	NIEM-T10-DiffDC	IF Count(XElm(itl:PackageFacialAndSMTImageRecord)) GT 1 THEN ForEach(Pair (A,B) of XElm(itl:PackageFacialAndSMTImageRecord)		X

Field	Section	Description	Num	M/O/D	Mnemonic	Rule / Content	Test	Code
						{ IF { XElm(nc:IdentificationID) in XElm(biom:PhysicalFeatureReferenceIdentificati on) in A} EQ { XElm(nc:IdentificationID) in XElm(biom:PhysicalFeatureReferenceIdentificati on) in B} THEN { XElm(nc:IdentificationID) in XElm(biom:ImageReferenceIdentification) in A} EQ { XElm(nc:IdentificationID) in XElm(biom:ImageReferenceIdentification) in B} }		B
			3	D	10.039-T10-SameBodyP art	<Unsupported: Not feasible to test if the images are related to the same part of the body>	t-1	B
Field: 14.026-SimCap	7.3, 7.3.5	In order to accommodate the emergence of technology that can simultaneously capture fingerprint images on separate platens or other technology that does not preserve the full relative position of the fingers to each other, Field 14.026: Simultaneous capture / SCF allows the user to specify the same reference number for all images that were simultaneously captured.	3	O	14.026-SCF	<Unsupported: Not feasible to test if the images were captured simultaneously. >	t-1	B
Field: xx.902-ANN	7.4.1	New for this version of the standard optional field xx.902 is used to store annotation, logging, or processing information associated with one or more processing algorithms or workstations. If present, this text field shall consist of one or more subfields comprised of a set of information items. Four mandatory information items comprise a subfield: • The first information item is the GMT date and time / GMT when the processing occurred. (See Section 7.7 2.2) • The second information item (processing algorithm name/version / NAV) shall contain text of up to 64 characters identifying the name and version of the processing algorithm/application or workstation. • The third information item (algorithm	1	O	xx.902-SubfieldCo unt	Count(Subfields in Record.902) GTE 1		B*
			1	O	xx.902-InfoItemCo unt	ForEach(Subfield in Record.902) { Count(InfoItems in Subfield) EQ 4 }		B*
			1	M ⇑	xx.902-GMT-Value	ForEach(Subfield in Record.902) {InfoItem:1 in Subfield} MO [ValidUTC/GMT]	t-6	T
			1	M ⇑	NIEM-xx.902-ANN-GMT	<The treatment of subfields for validation in the XML version requires further review.>	t-6	X*
			1	M	xx.902-	ForEach(Subfield in Record.902)		B*

Field	Section	Description	Count	M/O	Code	Format		Type
		owner / OWN) shall contain text of up to 64 characters with the contact information for the organization that developed/maintains the processing algorithm/application or latent workstation. • The fourth and final information item (process description / PRO) shall contain text of up to 255 characters describing a process or procedure applied to the sample in this Type-XX record.		⇐	NAV-CharCount	{ Length(InfoItem:2 in Subfield) MO [1 to 64] }		B*
			1	M ⇐	xx.902-OWN-CharCount	ForEach(Subfield in Record.902) { Length(InfoItem:3 in Subfield) MO [1 to 64] }		
			1	M ⇐	xx.902-PRO-CharCount	ForEach(Subfield in Record.902) { Length(InfoItem:4 in Subfield) MO [1 to 255] }		B*
Field: 9.901-ULA	7.4.2	This optional field, which is new to this version of the standard, exists only in Record Type-9. The ULW has been extensively used and logs generated from it were routinely transmitted in user-defined Field 9.901 in previous versions of this standard. Thus, this version of the standard formally includes Field 9.901: Universal latent workstation annotation information / ULA to record latent processing logs formatted according to the ULW.		O	9.901-ULW	<The test assertions for this type may not be supported in this version of the CTM. If they are supported, they are included under field testing for Record Type-9: Minutiae data record.>	t-2	
Field: 98.900-ALF	7.4.3	If a user wishes to maintain a log of differences between transmissions,... may be used to indicate how and why a transaction was modified. Record Type-98 is new to this version of the standard.		O	98.900-ALF	<The test assertions for this type may not be supported in this version of the CTM. If they are supported, they are included under field testing for Record Type-98: Information assurance record.>	t-2	
Field: Comment	7.4.4	The optional Comment field appears in many record types and may be used to insert free text information. It is not reserved exclusively for log-related information but has historically often been used for this purpose. It is limited to a maximum of 126 characters. The comment fields are: Field 10.038 Field 13.020 Field 14.020 Field 15.020 Field 16.020 Field 17.021 Field 18.022 Field 19.020	1	O	10.038-CharCount	Length(10.038) LTE 126		B
			1	O	xx.020-CharCount	Length([13,14, 15],020) MO [1 to 126]		B
			1	O	16.020-CharCount	Length(16.020) MO [1 to 126]		B
			1	O	17.021-CharCount	Length(17.021) MO [1 to 126]		B

	Field 20.020 Field 21.020 The EFS comment field in Record Type-9 is limited to 200 characters. It is: Field 9.351: EFS comments / COM	1	O	18.022-CharCount	Length(18.022) MO [1 to126]	B
		1	O	19.020-CharCount	Length(19.020) MO [1 to126]	B
		1	O	20.020-CharCount	Length(20.020) MO [1 to126]	B
		1	O	21.020--CharCount	Length(21.020) MO [1 to126]	B
Field: 98.003-IA/DFO	7.5.1	The Record Type-98: Information assurance record, which is new to this version of the standard, allows special data protection procedures to ensure the integrity of the transmitted data. Field 98.003: IA data format owner / DFO and Field 98.005: IA data format type / DFT define the information assurance regime that is employed to store data in Fields 98.200-899: User-defined fields / UDF.		M	98.003-DFO	<The test assertions for this type may not be supported in this version of the CTM. If they are supported, they are included under field testing for Record Type-98: Information assurance record.>	t-2

Field: HAS	7.5.2	Optional field xx.996, which is new to this version of the standard, is designed for use in Record types 10 and above that have a Field xx.999 storing the biometric data. It is comprised of 64 characters	1	O	xx.996 - RecordTypes	ForEach(Field ST FieldNumber(Field) EQ 996)
						{
						Type(ParentRecord(Field)) GTE 10
						}
							T

			1	O	NIEM-HAS	ForEach(Field ST Field EQ	X

31

Description	No.		Field	Cnt	Assertion		
			RecordTypes		XElm(biom:ImageHashValue)) { XElm(biom:RecordCategoryCode) in ParentRecord(Field)) GTE 10 }		B
representing hexadecimal values. Thus, each character may be a digit from "0" to "9" or a letter "A" through "F".		1 O	xx.996-CharType H	1	ForEach(Field ST FieldNumber(Field) EQ 996) { Bytes(Field) MO[ASCII(0 to 9, A to F)] }		B
		1 O	xx.996-CharCount	1	ForEach(Field ST FieldNumber(Field) EQ 996) { Length(Field) EQ 64 }		B
In the 2007 version of the standard, Record Type-1 fields for agency identification were comprised of one information item {destination}{originating} agency identifier / DAI or ORI. The 2008 version of the standard added a second optional information item {destination}{originating} agency name / DAN or OAN, and is a text description of the organization name.	7.6 Field: Agency Codes	1 M	Fields: DAI/ORI		\<See Record Type-1 assertions associated with Fields 1.007 and 1.008 in "8.1: Record Type-1: Transaction Information Record".\>	t-2	B
In this version of the standard, the agency names (DAN and OAN) are contained in a new field (Field 1.017 Agency names / ANM) since information items cannot be added to existing fields in Traditional encoding and still preserve backward compatibility. DAN and OAN have an unlimited maximum number of characters in this version. XML encoding is not dependent upon the field number, so there is no change required for compatibility with the 2008 version. Both information items in ANM are optional and may be encoded using alphanumeric characters with any special characters allowed in ASCII. The affected fields are: → Field 1.007 Destination agency identifier / DAI → Field 1.008 Originating agency		1 M	Fields: DAN/OAN		\<See Record Type-1 assertions associated with Field 1.017 in "8.1: Record Type-1: Transaction Information Record".\>	t-2	B

Field	Section	Description		M/O		Name	Condition		B
		identifier / ORI → Field 1.017 Agency names / ANM							B
Field: Originating Agency	7.6, Table 22, Table 58, Table 70, Table 71, Table 75	In many Record types, Field xx.004 contains the SRC. This is the identifier of the agency that actually created the record and supplied the information contained in it…SRC is unlimited in size and is "U" character type.	1	M	Fields-SRC Length		Length([10,13 to 21, 98, 99].004) GTE 1		B
			1	M	[10,13 to 21, 98, 99].004-Value		TRUE		
Field: Source Agency Name	7.6, Table 22, Table 58, Table 70, Table 71, Table 75	In order to maintain backward compatibility with the 2007 version while maintaining backward compatibility with the 2008 version, a new optional Field xx.993 has been added for the Source agency name / SAN. SAN is up to 125 characters and in "U" character type.	1	M	Fields-SAN Length		ForEach(Field ST FieldNumber(Field) EQ 993) { Length(Field) MO [1 to 125] }		B
			1	M	xx.993-Value		TRUE		
Field: Device ID	7.7.1.1	The DUI shall contain a string uniquely identifying the device or source of the data. This field shall be one of: • Host MAC address, identified by the first character "M", or • Host processor ID, identified by the first character "P". Fields containing the DUI are: -Field 9.903: Device unique identifier / DUI -Field 10.903: Device unique identifier / DUI -Field 13.903: Device unique identifier / DUI -Field 14.903: Device unique identifier / DUI -Field 15.903: Device unique identifier / DUI -Field 16.903: Device unique identifier / DUI -Field 17.017: Device unique identifier / DUI -Field 19.903: Device unique identifier / DUI -Field 20.903: Device unique identifier / DUI -Field 99.903: Device unique identifier / DUI The MAC address takes the form of six	1	O	xx.903, 17.017-Value		IF(First(Byte in [9,10,13 to 16, 19, 20, 99].903, 17.017) EQ ASCII(M) THEN Bytes([9,10,13 to 16, 19, 20, 99].903) MO[ASCII|0 to 9, A to F]) AND Length([9,10,13 to 16, 19, 20, 99].903) EQ 13 } IF(First(Byte in [9,10,13 to 16, 19, 20, 99].903) EQ ASCII(P) THEN { Length([10,13,14].903) EQ MO[13 to 16] }		B

33

Field	Section	Count		Code	Condition	Type
					pairs of hexadecimal values (0 through 9 and A through F). They are represented without separators in this standard for a total of 13 characters. The processor ID may be up to 16 characters.	
Field: Make Model	7.7.1.2	1	O	xx.904-SubfieldCount	The MMS contains the make, model and serial number for the capture device. It shall consist of three information items. Each information item shall be 1 to 50 characters. Any or all information items may indicate that information is unknown with the value "0". Fields containing the MMS are: -Field 9.904: Make/model/serial number -Field 10.904: Make/model/serial number -Field 13.904: Make/model/serial number -Field 14.904: Make/model/serial number -Field 15.904: Make/model/serial number -Field 16.904: Make/model/serial number -Field 17.019: Make/model/serial number -Field 19.904: Make/model/serial number -Field 20.904: Make/model/serial number -Field 99.904: Make/model/serial number Count(InfoItems in [9, 10,13 to16, 19, 20, 99].904) EQ 3	B*
		1	O	xx.904-[MAK, MOD, SER]-CharCount	Length(InfoItems in [9, 10,13 to 16, 19, 20, 99].904) LTE 50 AND GTE 1	B*
		1	O	17.019-SubfieldCount	Count(InfoItems in 17.019) EQ 3	B*
		1	O	17.019-[MAK, MOD, SER]-CharCount	Length(InfoItems in 17.019) LTE 50 AND GTE 1	B*
Field: Device Monitoring	7.7.1.3, Table 5	1	O	xx.030-Value	This field describes the level of human monitoring that was associated with the biometric sample capture. Alphabetic values are selected from Table 5. These are corresponding fields in the standard: -Field 10.030: Device monitoring mode / DMM -Field 14.030: Device monitoring mode / DMM -Field 15.030: Device monitoring mode / DMM -Field 16.030: Device monitoring mode / DMM -Field 17.030: Device monitoring mode / DMM -Field 19.030: Device monitoring mode / DMM {[10,14 to 17, 19].030} MO [ASCII(CONTROLLED, ASSISTED, OBSERVED, UNATTENDED, UNKNOWN)]	B

Field	Reference	Description		M/D/O	Fields	Assertion		Code
Field: UTC	7.7.2.2	UTC has replaced GMT as the main reference time scale terminology, but the older terminology is retained in this standard for existing record types. In this standard, Field 1.014 Greenwich mean time / GMT shall be taken to mean the UTC value. Some newer record types using this format refer to the data as UTC (such as in Field 18.013: Sample collection date / SCD). This time is independent of the actual time zone where the time and date s recorded. The data is YYYYMMDDhhmmssZ, where the Z is indicates the zone description of 0 hours.		M	Fields-UTC	\<This requirement applies to any field that contains a date or time value. The assertions will be applied individually for those fields.\>	t-2, t-6	
Field: TIX	7.7.2.5, Table 86, Table 89	For Type-20 or Type-21 records containing video or audio, this field shall contain two information items, time index start / TIS and time index end / TIE for the start and end times of segments within a video or audio file, measured in hh:mm:ss.sss where ss.sss refers to the seconds and milliseconds. Thus, the allowed special characters are the colon and the period. TIX is comprised of one or more subfields. Each subfield corresponds to a single segment, with a starting and end time as separate information items.		D	Fields-TIX	\<The test assertions for this type may not be supported in this version of the CTM. If they are supported, they are included under field testing for Record Type-98: Information assurance record.\>	t-2	
Field: Geographic	7.7.3, Table 57, Table 70, Table 71, Table 73, Table 74, Table 75, Table 79, Table 85, Table 86, Table 91	New to this version of the standard, this optional field (xx.998) is used in most Record Types 10 and above. It specifies the coordinated universal time (UTC) and the location where the biometric sample was collected. All of this information is contained in up to fifteen information items.	1	O	xx.998-RecordTypes	ForEach(Field ST FieldNumber(Field) EQ 998) { Type(ParentRecord(Field)) GTE 10 }		T
			1	O	NIEM-GEO-RecordTypes	ForEach(Field ST Field EQ XElm(biom:CaptureLocation)) { (XElm(biom:RecordCategoryCode) in ParentRecord(Field)) GTE 10 }		X
			1	O	xx.998-InfoItemCount	ForEach(Field ST FieldNumber(Field) EQ 998) { Count(InfoItems in Field) LTE 15 }		B*
			1	0	xx.998-	ForEach(Field ST FieldNumber(Field) EQ 998)		

		Field	Condition						
1	O	SubfieldCount xx.998-[UTE, LTD, LTM, LTS, LGD, LGM, LGS, ELE, GDC, GCM, GCE, GCN ']-CharType	{ Count(Subfields in Field) EQ 1 } ForEach(Field ST FieldNumber(Field) EQ 998) { Bytes(InfoItem:1 in Field)) MO [0x30 to 0x39, 0x5A] AND Bytes(InfoItem:2,5,8 in Field) MO [0x2D, 0x2E, 0x30 to 0x39] AND Bytes(InfoItem:3,4,6,7 in Field) MO [0x2E, 0x30 to 0x39] AND Bytes(InfoItem:11,12 Field) MO [0x30 to 0x39] AND Bytes(InfoItem:9,10 in Field) MO [0x30 to 0x39, 0x20, 0x41 to 0x5A, 0x61 to 0x7A]}						T
1	O	xx.998-[GRT, OSI, OCV]-CharType	TRUE						T
1	O	NIEM-xx.998-Subfield CharType	< The treatment of subfields for validation in the XML version requires further review. Byte values allowed for first "subfield" in XML are 0x30 to 0x39, 0x3A, 0x54, 0x5A.>						X*
1	O	xx.998-[UTE, LTD, LTM, LTS, LGD, LGM, LGS, ELE, GDC, GCM, GCE, GCN, GRT, OSI, OCV]-CharCount	ForEach(Field ST FieldNumber(Field) EQ 998) { Length(InfoItem:1 in Field)) EQ 15 AND Length(InfoItem:2 in Field) MO [1 to 9] AND Length(InfoItem:3,4,6 to 8,12 in Field) MO [1 to 8] AND Length(InfoItem:5,14 in Field) MO [1 to 10] AND Length(InfoItem:9 in Field) MO [3 to 6] AND Length(InfoItem:10 in Field) MO [2, 3] AND Length(InfoItem:11 in Field) MO [1 to 6] AND Length(InfoItem:13 in Field) MO [1 to 150] AND Length(InfoItem:15 in Field) MO [1 to 126]						T

Section	Field	Description	#	O	Element	Condition	Length	Type
						}		X*
			1	O	NIEM-xx.998-SubfieldCharCount	< The treatment of subfields for validation in the XML version requires further review. Length of the first "subfield" in XML is 20.>		X*
7.7.3	Field: Geographic-Subfield 1	The first information item is optional. It is the coordinated universal time entry /UTE. See Section 7.7.2.2.	1	O ⇑	xx.998-UTE-Value	ForEach(Field ST FieldNumber(Field) EQ 998) {Infoltem:1 in Field} MO [ValidUTC/GMT]	t-6	T
			1	O ⇑	NIEM-xx.998-UTE-Value	< The treatment of subfields for validation in the XML version requires further review.>	t-6	X*
7.7.3	Field: Geographic-Conditional	The next eight information items (information items 2 through 9) comprise the Geographic Coordinate Latitude/Longitude. As a group, they are optional. However, latitude degree value / LTD and longitude degree value / LGD are co-conditional, so they shall both be present if either is present. Further, "minutes" values LTM and LGM can only be present if their corresponding	2	-	xx.998-LTD-LGD-Conditional	ForEach(Field ST FieldNumber(Field) EQ 998) { Present(Infoltem:2 in Field} IFF Present (Infoltem:5 in Field} }		B*
		"degrees" values are present. LTS and LGS can only be present if their corresponding "minutes" value is present. The other entries are optional. If a decimal value is used in a particular information item, the more granular information item shall be empty (e.g., if Longitude minutes equals 45 27, Longitude seconds shall be empty).	2	-	xx.998-LTM-LGM-Conditional	ForEach(Field ST FieldNumber(Field) EQ 998) { Present(Infoltem:3 in Field) IFF Present (Infoltem:6 in Field) }		B*
			2	-	xx.998-LTS-LGS-Conditional	ForEach(Field ST FieldNumber(Field) EQ 998) { Present(Infoltem:4 in Field) IFF Present (Infoltem:7 in Field) }		B*
		LTM and LGM are co-conditional, so they shall both be present if either is present.	2	-	xx.998-GCM-GCE-GCN-Conditional	ForEach(Field ST FieldNumber(Field) EQ 998) { IF Present(Infoltem:10 OR Infoltem:11 OR Infoltem:12 in Field) THEN Present(Infoltem:10 AND Infoltem:11 AND Infoltem:12 in Field) }		B*
		LTS and LGS are co-conditional, so they shall both be present if either is present. If LTM is present, then LGM shall be present. If LTS is present, then LGS shall be present.	2	O	xx.998-LTS-LTM-Conditional	ForEach(Field ST FieldNumber(Field) EQ 998) { IF Present (Infoltem:4 in Field) THEN Present(Infoltem:3 in Field) }		B*
		The tenth, eleventh and twelfth information items are treated as a group and are optional.	2	O	xx.998-LTM-LTD-Conditional	ForEach(Field ST FieldNumber(Field) EQ 998) { IF Present(Infoltem:3 in Field) THEN Present (Infoltem:2 in Field) }		B*
		These three information items together	2	O	xx.998-LGS-	ForEach(Field ST FieldNumber(Field) EQ 998)		B*

Section	Description	Count	O	Name	Condition						B*
	are a coordinate which represents a location with a Universal Transverse Mercator (UTM) coordinate. If any of these three information items is present, all shall be present. A fifteenth optional information item is the geographic coordinate other system value / OCV. It shall only be present if OSI is present in the record.			LGM-Conditional	{ IF Present(InfoItem:7 in Field) THEN Present (InfoItem:6 in Field) }						B*
		2	O	xx.998-LGM-LGD-Conditional	ForEach(Field ST FieldNumber(Field) EQ 998) { IF Present(InfoItem:6 in Field) THEN Present (InfoItem:5 in Field) }						B*
		2	O	xx.998-LGD-LGM-LGS-Conditional	ForEach(Field ST FieldNumber(Field) EQ 998) { IF {InfoItem:2 in Field} MOD 1 NEQ 0 THEN Length(InfoItem:3 in Field) EQ 0 AND Length(InfoItem:4) EQ 0 AND IF {InfoItem:3 in Field} MOD 1 NEQ 0 THEN Length(InfoItem:4 in Field) EQ 0 }						B*
		2	O	xx.998-LTD-LTM-LTS-Conditional	ForEach(Field ST FieldNumber(Field) EQ 998) { IF {InfoItem:5 in Field} MOD 1 NEQ 0 THEN Length(InfoItem:6 in Field) EQ 0 AND Length(InfoItem:7) EQ 0 AND IF {InfoItem:6 in Field} MOD 1 NEQ 0 THEN Length(InfoItem:7 in Field) EQ 0 }						B*
		2	O	xx.998-OCV-OSI-Conditional	ForEach(Field ST FieldNumber(Field) EQ 998) { IF Present(InfoItem:14 in Field) THEN Present (InfoItem:13 in Field) }						B*
7.7.3 Field: Geographic-Values-SubfieldCount 2 to 8	The second information item is latitude degree value / LTD. This is a value that specifies the degree of latitude. The value shall be between -90 (inclusive) and +90 (inclusive). The third information item is latitude minute value / LTM. This is a value that specifies a minute of a degree. The value shall be between 0 (inclusive) to 60 (exclusive).	1	-	xx.998-[LTD, LTM, LTS, LGD, LGM, LGS, ELE]-Value	ForEach(Field ST FieldNumber(Field) EQ 998) { {InfoItem:2 in Field} GTE -90 AND LTE 90 AND {InfoItem:3 in Field} GTE 0 AND LT 60 AND						B*

38

Field	Ref	Description	Min	Value Name	Max	Condition	Char Type	Note
		The fourth information item is the latitude second value / LTS. This is a value that specifies a second of a minute. The integer shall be 0 (inclusive) to 60 (exclusive).				{InfoItem:4 in Field} GT 0 AND LT 60 AND		
		The fifth information item is the longitude degree value / LGD. It is a value that specifies the degree of a longitude. The value shall be between -180 (inclusive) and +180 (inclusive).				{InfoItem:5 in Field} GTE -180 AND LTE 180 AND		
		The sixth information item is the longitude minute value / LGM. It is a value that specifies a minute of a degree. The value shall be from 0 (inclusive) to 60 (exclusive).				{InfoItem:6 in Field} GTE 0 AND LT 60 AND		
		The seventh information item is the longitude second value / LGS. This is a value that specifies a second of a minute. The integer shall be 0 (inclusive) to 60 (exclusive).				{InfoItem:7 in Field} GT 0 AND LT 60 AND		
		The eighth information item is the elevation / ELE. It is expressed in meters. It is a numeric value. It is between -422 meters (Dead Sea) and 8848 meters (Mount Everest).				{InfoItem:8 in Field} GT -422 AND LT 8848 }		
Field: Geographic-Values-SubField 9	7.7.3, Table 6	The ninth information item is the geodetic datum code / GDC. It is an alphanumeric value of 3 to 6 characters in length. This information item is used to indicate which coordinate system was used to represent the values in information items 2 through 7. If no entry is made in this information item, then the basis for the values entered in the first eight information items shall be WGS84, the code for the World Geodetic Survey 1984 version - WGS 84 (G873). See Table 6 for values.	1	xx.998-GDC-Value	O	ForEach(Field ST FieldNumber(Field) EQ 998) { {InfoItem:9 in Field} MO [ASCII(AIRY, AUST, BES, BESN, CLK66, CLK80, EVER, FIS60, FIS68, GRS67, HELM, HOUG, INT, KRAS, AIRYM, EVERM, FIS60M, SA69, WGS60, WGS66, WGS72, WGS84)] OR Length(InfoItem:9 in Field) MO [3 to 6] AND Bytes(InfoItem:9 in Field) MO [0x30 to 0x39, 0x20, 0x41 to 0x5A, 0x61 to 0x7A] }	B*	t-7
Field: Geographic-Values-SubField 10	7.7.3	The tenth information item is the geographic coordinate universal transverse Mercator zone / GCM. It is an alphanumeric value of 2 to 3 characters.	1	xx.998-GCM-Value	O	ForEach(Field ST FieldNumber(Field) EQ 998) { IF Length(InfoItem:10 in Field) EQ 2	B*	t-8

39

Field: Geographic-Values- SubField 11 to 15	7.7.3						B*

		This is a one or two digit UTM zone number followed by the 8 degree latitudinal band designator (which is a single letter). Valid latitudinal band designators include C through X, omitting I and O.	THEN
			{Bytes:1 in InfoItem:10 in Field} MO [1 to 9]
			AND
			
			{Bytes:2 in InfoItem:10 in Field} MO [ASCII(C to X)] AND NOT MO [ASCII(I,O)]
			
			ELSE
			{Bytes:1,2 in InfoItem:10 in Field} MO [10 to 60]
			
			AND
			
			{Bytes:3 in InfoItem:10 in Field} MO [ASCII(C to X)] AND NOT MO [ASCII(I,O)]
			
			}				

		The eleventh information item is the geographic coordinate universal transverse Mercator easting / GCE. It is an integer of 1 to 6 digits. The twelfth information item is the geographic coordinate universal transverse Mercator northing / GCN. It is an integer of 1 to 8 digits. The thirteenth information item is optional. It is the geographic reference text / GRT. This information item is an alphanumeric entry of up to 150 characters. It is a free form text describing a street address or other physical location (such as a band of Washington and Madison, Geneva, NYison, Geneva, NYNYdress or other physical lgeographic coordinate other system identifier / OSI allows for other coordinate systems. This information item specifies the system identifier. It is up to 10 characters in length. A fifteenth optional information item is the	2	D	xx.998- [GCE, GCN, GRT, OSI, OCV]-Value	TRUE	

Field	Section	Num	M/D	Field ID	Description	Test Assertion	t	Status
					geographic coordinate other system value / OCV. It shall only be present if OSI is present in the record. It can be up to 126 characters in length.			
Field: Impression-Values	7.7.4.1, Table 7, 8.4.3, 8.9.3, 8.13.3, 8.14.3, 8.15.3, 8.19.3		M	[4,9,13,14,15,19].003-IMP	This field contains a code from Table 7 for how the friction ridge sample was collected. It has been expanded in this version of the standard to include plantars and unknowns.	<The test assertions for these types vary for this requirement, and therefore included under field testing for the specific Record Types.>	t-2	
Field: FGP-Values	7.7.4.2, Table 8, 8.4.4, 8.9.5.9, 8.13.13, 8.14.13, 8.15.13, 8.19.13		M	4.004, 9.134, [13,14,15,19].013-FGP	FGP is used in Record types dealing with friction ridges. It specifies which friction ridge biometric sample was collected. Note that for codes 1-40 and 60-84, the Table 8 specifies recommended MAXIMUM width and height. (Individual implementation domains and application profiles may use different values.) In previous versions of this standard, FGP was used for finger position, and PLP for palmprint position. They are now in one table, along with the codes added in the ANSI/NIST-ITL 1a-2009 amendment. New to this version, plantar codes are included in the table. In order to cover all of these cases, the name was changed to friction ridge generalized position / FGP.	<The test assertions for these types vary for this requirement, and therefore included under field testing for the specific Record Types.>	t-2	
Field: PPD Conditional	7.7.4.3	2	D	14.013-14.014-Conditional	For exemplar fingerprints contained in Type-14 records, if the impression is known to be an entire joint image (EJI), full finger view (FFV), or extreme tip (TIP), then Field 14.013: Friction ridge generalized position / FGP shall be set to 19, and Field 14.014: Print position descriptors / PPD shall be specified; Field 14.015: Print position coordinates /PPC may be (optionally) specified.	Present(14.014) IFF Present(InfoItem in 14.013 ST {InfoItem} EQ 19)		B*
		2	D	14.013-14.015-Conditional		IF Present(14.015) THEN Present(InfoItem in 14.013 ST {InfoItem} EQ 19)		B*
Field: SPD,PPC Conditional	7.7.4.3	2	D	13.013-13.014-Conditional	For latent prints contained in Type-13 records, if all or part of the impression should be compared against the medial or proximal segments or the extreme tips, then Field 13.013: Friction ridge generalized position / FGP shall be set to	Present(13.014) IFF Present(InfoItem in 13.013 ST {InfoItem} EQ 19)		B*
		2	D	13.013-13.015-Conditional		IF Present(13.015) THEN Present(InfoItem in 13.013 ST {InfoItem} EQ 19)		B*

Field	Reference	Description	Count	M/O/D	Subfield	Condition		B*
		19, and Field 13.014: Search position descriptors / SPD shall be specified; Field 13.015: Print position coordinates / PPC may be (optionally) specified.						
Field: SPD, PPD Values	7.7.4.3, Table 8, Table 9	The position descriptor, in Field 13.014: Search position descriptors / SPD or Field 14.014: Print position descriptors / PPD contains two mandatory information items: For a Type-13 record (latent prints), the first information item (probable decimal finger position code / PDF) (0-10, 16 or 17) is taken from Table 8. A "0" indicates that all the fingers of a possible candidate should be searched. For a Type-14 record (known exemplars), the first information item is the decimal finger position code / DFP. It is also taken from Table 8 with a value of 1 to 10, inclusive or 16 or 17. The second information item (finger image code / FIC) is the code taken from Table 9 to indicate the portion of the database to search. Full-length finger joint images use codes FV1 through FV4. Figure 4 is an illustration of the Entire Joint Image for a middle finger with each of the full finger views and constituent parts identified. Multiple portions of the EJI may be listed in a separate subfield.	1	D	13.014-PDF-Value	ForEach(Subfield in 13.014) { InfoItem:1 in Subfield) MO [0 to 10, 16,17] AND MO [Integers] }		B*
			1	D	14.014-DFP-Value	ForEach(Subfield in 14.014) { InfoItem:1 in Subfield) MO [1 to 10, 16,17] }		B*
			1	D	[13,14].014-FIC-Value	ForEach(Subfield in [13,14].014) { InfoItem:2 in Subfield) MO [ASCII(EJI, TIP, FV1, FV2, FV3, FV4, PRX, DST, MED)] }		B*
Field: PPC-Subfield Occurrences	7.7.4.4, Table 9, Table 70, Table 71	When used, Field 13.015: Print position coordinates / PPC or Field 14.015: Print position coordinates / PPC shall consist of six (6) mandatory information items describing the type or portion of the image contained in this record and its location within an EJI... Individual full finger or segment definitions may be repeated as repeating sets of information items.	1	O	[13,14].015-SubfieldCount	Count(Subfields in [13,14].015) MO [1 to 12]		B*
			1	O	[13,14].015-InfoItemCount	ForEach(Subfield in [13,14].015) { Count(InfoItems in Subfield) EQ 6 }		B*
Field: PPC-Subfield 1	7.7.4.4, Table 9, Table 70, Table 71	The first information item is the full finger view / FVC with values of "FV1" through "FV4". Values of "FV1" to "FV4" specify the perspective for each full finger view. For the case of a fingertip, the first	1	M ⇑	[13,14].015-FVC-Value	ForEach(Subfield in [13,14].015) { InfoItem:1 in Subfield) MO [ASCII(FV1, FV2, FV3, FV4, TIP, NA)] }		B*

Field	Description	Reference	#	Char	Field Value	Condition	Status
	information item shall be "TIP". FVC will contain the code "NA" if only a proximal, distal or medial segment is available.						B*
Field: PPC-Subfield 2	The second information item is used to identify the location of a segment / LOS within a full finger view. LOS will contain the *not applicable* code "NA" if the image portion refers to a full finger view, tip or to the entire joint image locations. Otherwise, it shall contain "PRX", "DST", "MED" for a proximal, distal, or medial segment, respectively.	7.7.4.4, Table 9, Table 70, Table 71	1	M ⇑	[13,14].015 - LOS-Value	ForEach(Subfield in [13,14].015) { [InfoItem:2 in Subfield] MO [ASCII(PRX, DST, MED, NA)] }	B*
Field: PPC-SubfieldCount 3,4	The third information item is the left horizontal coordinate / LHC. It is the horizontal offset in pixels to the left edge of the bounding box relative to the origin positioned in the upper left corner of the image.	7.7.4.4, Table 9, Table 70, Table 71	2	M ⇑	[13,14].015-LHC-Value	ForEach(Subfield in [13,14].015) { [InfoItem:3 in Subfield] GTE 0 AND LTE {[13,14].006} AND MO [Integers] }	B*
	The fourth information item is the right horizontal coordinate / RHC. It is the horizontal offset in pixels to the right edge of the bounding box relative to the origin positioned in the upper left corner of the image.		2	M ⇑	[13,14].015-RHC-Value	ForEach(Subfield in [13,14].015) { [InfoItem:4 in Subfield] GTE {InfoItem:3 in Subfield} AND LTE {[13,14].006} AND MO [Integers] }	B*
Field: PPC-SubfieldCount 5,6	The fifth information item is the top vertical coordinate / TVC is the vertical offset (pixel counts down) to the top of the bounding box.	7.7.4.4, Table 9, Table 70, Table 71	2	M ⇑	[13,14].015-TVC-Value	ForEach(Subfield in [13,14].015) { [InfoItem:5 in Subfield] GTE 0 AND LTE {[13,14].007} AND MO [Integers] }	B*
	The sixth information item is the bottom vertical coordinate / BVC. It is the vertical offset from the upper left corner of the image down to the bottom of the bounding box. It is counted in pixels.		2	M ⇑	[13,14].015-BVC-Value	ForEach(Subfield in [13,14].015) { [InfoItem:6 in Subfield] GTE {InfoItem:5 in Subfield} AND LTE {[13,14].007} AND MO [Integers] }	B*
Field: SAP Conditional	SAP codes are mandatory in Type-10 records with a face image. FAP is optional in Type-14. IAP is optional in Type-17 records. The Subject Acquisition Profile (SAP) is a mandatory field when Field 10.003: Image type / IMT contains "FACE". Otherwise, it shall not be entered.	7.7.5, 8.10.13	2	D	10.003-10.013-Conditional	Present(10.013) IFF {10.003} EQ ASCII (FACE)	B

43

Name	Section	Description	#	M/O/D	Field	Condition/Value	Test	Cat
Field: SAP Values	7.7.5.1, Table 10	Field 10.013: Subject acquisition profile / SAP has the SAP level code for face in Table 10. The SAP codes 32, 42 and 52 are new to this version of the standard.	1	D	10.013-Values	{10.013} MO [0,1,10-15,20,30,32,40,42,50-52]		B
Field: SAP-Level Requirements	7.7.5.1.1 to 7.7.5.1.10, Table 11, Annex E	<Sections 7.7.5.1.1 to 7.7.5.1.10 describe requirements for the image for various SAP Levels>	3	D	10.013-Levels	<Unsupported: Determining the condition under which the image was captured to verify the SAP Level is not feasible at this time.>	t-9	
Field: FAP Values	7.7.5.2, Table 12	The profile levels for fingerprint acquisition are optional and are based upon those listed in the *Mobile ID Best Practice Recommendation*. They are entered in Field 14.031: Subject acquisition profile – fingerprint / FAP, which is new to this version of the standard.	1	O	14.031-Values	{14.031} MO [10,20,30,40,50,60]		B
Field: FAP-Level Requirements	7.7.5.2, Table 12	<Section 7.7.5.2 and Table 12 describe requirements for the image for various FAP Levels>	3	O	14.031-Levels	<Unsupported: Determining the condition under which the image was captured to verify the FAP Level is not feasible at this time.>	t-9	
Field: IAP Values	7.7.5.3, Table 75	The profile levels for iris acquisition, which are new to this version of the standard, are optional and are based upon those listed in the *Mobile ID Best Practice Recommendation (BPR)* (See Annex G: Bibliography). They are entered in Field 17.031: Subject acquisition profile – iris / IAP.	1	O	17.031-Values	{17.031} MO [20,30,40]		B
Field: IAP-Level Requirements	7.7.5.3	<Section 7.7.5.3 describes requirements for the image for various IAP Levels.>	3	M	17.031-IAP Levels	<Unsupported: Determining the condition under which the image was captured to verify the IAP Level is not feasible at this time.>	t-9	
Field: Image-Coordinates	7.7.6	Each image formatted in accordance with this standard shall appear to have been captured in an upright position and approximately centered horizontally in the field of view...The y-coordinate (vertical) position shall increase positively from the origin to the bottom of the image.	3	M	Images-Coordinates	<Not directly tested, but this convention is used when testing for conformance in each of the assertions related to images>	t-2	
Record: Type4-Resolution 500 Only	7.7.6	All record types containing images are variable resolution except for Type-4, which has a fixed resolution. Record Type-4 shall not be used for anything but the 500 ppi class.	2	M	[1.011, 1.012]-Conditional	IF(Present(Record ST Type(Record) EQ 4) THEN {1.011, 1.012} GTE 19 30 AND LTE 20.08 ELSE THEN		B

Field	Ref						
						{1.011, 1.012} EQ 00.00	
Field: Resolution Tolerance	7.7.6.1, 7.7.6.3, Table 14	1	M	Field-Resolution Tolerance	In this version, NSR and NTR only apply to Record Type-4: Grayscale fingerprint image… For Appendix F certified devices, resolution accuracy shall not vary more than 1% from the class resolution. A class resolution of 19.69 ppmm (500 ppi) has a lower bound of 19.49 ppmm (495ppi) and an upper bound of 19.89 ppmm (505ppi). For Personal Identity Verification (PIV) certified devices with fingerprint subject application profile (FAP) Levels 10 to 40 only resolution accuracy shall not vary more than 2% from the class Resolution. For example, a class resolution of 19.69 ppmm (500 ppi) has a lower bound of 19.30 ppmm (490ppi) and an upper bound of 20.08 ppmm (510ppi). Tolerance requirements shall apply to the class and nominal resolution requirements throughout this standard. This transmitting resolution does not have to be the same as the scanning resolution. However, the transmitting resolution shall be within the range of permissible resolution values for that record type.	\<This requirement specifies a tolerance for all resolution values expressed throughout the standard. If there is a resolution requirement for a type, the resolution must follow the 1% or 2% tolerance as described in the standard. This requirement will be applied to each applicable resolution requirement.\>	t-2
Field: NSR Conditional	7.7.6.2.1, Table 14	1	M	1.011-NSR-Length	If Type-4 records are included in the transaction, Field 1.011 Native scanning resolution / NSR contains five characters specifying the native scanning resolution in pixels per millimeter. It is expressed as two numeric characters followed by a decimal point and two more numeric characters (e.g. 19.69). This field is set to "00.00" if no Type-4 records are present in the transaction. With the deprecation of Record Types-3, 5 and 6, NSR only directly applies to Record Type-4 in this version of the standard. New to this version of the standard, NSR does not apply to Type-7	Length(1.011) EQ 5	B
		1	M	1.011-NSR-Value		Bytes:1,2,4,5 in 1.011 MO [0 to 9] AND Byte: 3 in 1.011 EQ "."	B

45

Field	Section	M/B	#	Field ID	Assertion		
					records, unless specified as such by an implementation domain.		
Field: Exemplar Scan Resolution	7.7.6.2.3				Record Type-14 shall be used if scanning a fingerprint image at the 1000 ppi class or above. It can also be used for the 500 ppi class.Record		
		M	2	Fields-Exemplar Valid Scan Resolutions	The migration path to higher scanning resolutions for image capturing devices with a native scanning resolution of the 500 ppi class shall be at a rate of 100% of the current native scanning resolution. Capture devices with native scanning resolutions not in step with this migration path shall provide (through subsampling, scaling, or interpolating downward) a nominal resolution that matches the next lower interval in the migration path. For example, a device with native scanning resolution of 47.24 ppmm (1200 ppi) shall provide a class resolution of 39.37 ppmm (1000 ppi)..	\<Scanning resolutions for Record Types 14,15,16,17,19 and 20 must migrate at a rate of 100% with a minimum of 500 ppi. These assertions are included under field testing for the specific Record Types>	t-2	B
Field: Latent Resolution	7.7.6.2.2, 7.7.6.1	M	2	13.008-Conditional	Latent images shall have a minimum class scanning resolution of 1000 ppi. ... resolution accuracy shall not vary more than 1% from the class resolution.	IF {13.008} EQ 1 THEN {13.016} AND {13.017} GTE 990 AND IF {13.008} EQ 2 THEN {13.016} AND {13.017} GTE 390		B
Field: Transmitting Resolution Required	7.7.6.3	M	2	Fields-Tx Resolution Required Type4	Each image to be exchanged shall have a specific resolution associated with the transmitted data. This transmitting resolution does not have to be the same as the scanning resolution. However, the transmitting resolution shall be within the range of permissible resolution values for that record type.	\<The test assertions are included under field testing for the respective record types >	t-2	
Field: Type4 NTR	7.7.6.3.1	M	1	1.012-CharCount	Field 1.012 Nominal resolution / NTR shall specify the transmitting resolution in pixels per millimeter. It is expressed as two numeric characters followed by a decimal point and two more numeric	Length(1.012) EQ 5		B
		M	1	1.012-Value	characters (e.g. 19.69). The transmitting resolution shall be within the	Bytes:1,2,4,5 in 1.012 MO [0 to 9] AND Byte: 3 in 1.012 EQ "."		B
		M	2	1.012-NTR 500 ppi	range 19.30 ppmm (490 ppi) to 20.08	\<See Requirement ID: "Record: Type4-Resolution 500 Only".>	t-2	B
		M	2	1.012-NTR LTE NSR	ppmm (510 ppi) for a Type-4 record . This	{1.012} LTE {1.011}		B

46 | | | | | | | |

46

						B*	
					t-2		

Section					Description
					range reflects the 2% tolerance from 500 ppi allowed for PIV certified devices. (See Table 14). For example, a senscr that scans natively at 508 ppi would list both NSR and NTR as 20 ppmm (= 508 ppi). These images should not be sampled down to exactly 500 ppi. This field is set to "00.00" if no Type-4 records are present in the transaction. Given that the transmitting resolution shall not be greater than the scanning resolution, images meant for identification applications, such as those from Appendix F certified devices (See Table 14) are restricted to a 1% deviance from 500 ppi.
7.7.6.3.2	Field: Variable Resolution THPS, TVPS	M	Fields-VarResolution	<The test assertions are included under field testing for the respective record types >	For variable-resolution friction ridge images (those in Record Types 13, 14, 15, 19 and possibly in Record Types 16 and 20), the transmitting resolution shall be at least as great as the class resolution of 500 ppi. There is no upper limit on the variable-resolution rate for transmission. However, the transmitting resolution shall not be greater than the scanning resolution. For variable resolution records, the Transmitted horizontal pixel scale / THPS and the Transmitted vertical pixel scale / TVPS shall be specified. (See Sections 7.7.8.4 and 7.7.8.5).
7.7.7	Field: Sample Quality Occurrences	1	-	[10, 13, 14, 15, 16, 17, 19].024, 9.135, 9..316, 14.023, 99.102- SubfieldCount	ForEach(Field in [9.135, 9.316, 10.024,13.024,14.023,14.024, 15.024, 16.024,17.024, 19.024, 99.102]) { Count(Subfields) MO [1 to 9] }

Many of the Record Types contain optional quality metric information. In addition to the three information items described here, a quality field may contain other information items. Each of the information items is contained in a subfield. Multiple subfields may be present, each indicating a different quality algorithm, up to a maximum of 9 times. Fields using this structure are:
-Field 9.135: M1 friction ridge quality data / FQD
-Field 9.316: EFS friction ridge quality metric / FQM
-Field 10.024: Subject quality score / SQS
-Field 13.024: Latent quality metric / LQM
-Field 14.023: Segmentation quality

		metric / SQM -Field 14.024: Fingerprint quality metric / FQM -Field 15.024: Friction ridge quality metric / FQM -Field 16.024: User-defined image quality metric / FQM -Field 17.024: Image quality score / IQS -Field 19.024: Friction ridge - plantar print quality metric / FQM -Field 99.102: Biometric data quality / BDQ									

Field: Sample Quality Subfield 1	7.7.7	The first information item shall be a quantitative expression of the predicted matching performance of the biometric sample, which is a quality value / QVU. This information item shall contain the integer image quality score between 0 and 100 (inclusive) assigned to the image data by a quality algorithm. Higher values indicate better quality. An entry of "255" shall indicate a failed attempt to calculate a quality score. An entry of
		"254" shall indicate that no attempt to calculate a quality score was made.
			1	-	[10, 13, 14, 15, 16, 17, 19].024, 9.135, 9..316, 14.023, 99.102-QVU-Value	ForEach(Field in
						[9.135, 9.316, 10.024, 16.024, 17.024, 99.102])
						{
						ForEach(Subfield in Field)
						{
						{InfoItem:1 in Subfield} MO [0 to 100, 254,255]
						AND MO [Integers]
						}
						}
						
						AND
						
						ForEach(Field in
						[13.024, 14.023, 14.024, 15.024, 19.024])
						{
						ForEach(Subfield in Field)
						{
						{InfoItem:2 in Subfield} MO [0 to 100, 254,255]
						AND MO [Integers]
						}
						}
											B*

Field: Sample Quality Subfield 2	7.7.7	A second information item shall specify the ID of the vendor of the quality algorithm used to calculate the quality score, which is an algorithm vendor identification / QAV. This 4-digit hex value (See Section 5.5 Character types) is assigned by IBIA and expressed as four characters. The IBIA maintains the Vendor Registry of CBEFF
		Biometric Organizations that map the value in this field to a registered organization.
			1	-	[10, 13, 14, 15, 16, 17, 19].024, 9.135, 9..316, 14.023, 99.102-QAV-Value	ForEach(Field in
						[9.135, 9.316, 10.024, 16.024, 17.024, 99.102])
						{
						
						ForEach(Subfield in Field)
						{
						{InfoItem:2 in Subfield} MO [0x0000 to 0xFFFF)]
						
						}
						}
						
						AND
						
						ForEach(Field in
							t-10				B*

				B*

1	-	[10, 13, 14, 15, 16, 17, 19].024, 9.135, 9..316, 14.023, 99.102-QAV-CharCount	[13.024, 14.023, 14.024, 15.024, 19.024])
			ForEach(Subfield in Field)
			{
			[InfoItem:3 in Subfield) MO [0x0000 to 0xFFFF]
			}
			}
			
			ForEach(Field in
			[9.135, 9.316, 10.024, 16.024, 17.024, 99.102])
			ForEach(Subfield in Field)
			{
			Length(InfoItem:2 in Subfield) EQ 4
			}
			}
			
			AND
			
			ForEach(Field in
			[13.024, 14.023, 14.024, 15.024, 19.024])
			ForEach(Subfield in Field)
			{
			Length(InfoItem:3 in Subfield) EQ 4
			}
			}	

			t-10	B*

1	-	[10, 13, 14, 15, 16, 17, 19].024, 9.135, 9..316, 14.023, 99.102-QAV-CharType	ForEach(Field in
			[9.135, 9.316, 10.024, 16.024, 17.024, 99.102])
			ForEach(Subfield in Field)
			{
			Bytes(InfoItem:2 in Subfield) MO [ASCII(A to F, 0 to 9)]
			}
			}
			
			AND
			
			ForEach(Field in
			[13.024, 14.023, 14.024, 15.024, 19.024])
			ForEach(Subfield in Field)
			{
			Bytes(InfoItem:3 in Subfield) MO [ASCII(A to F, 0 to 9)]
			}
			}	

Reference	Description			Value	Condition		B*
					}		
Field: Sample Quality Subfield 3							B*
7.7.7	A third information item shall specify a numeric product code assigned by the vendor of the quality algorithm, which may be registered with the IBIA, but registration is not required. This is the algorithm product identification / QAP that indicates which of the vendor's algorithms was used in the calculation of the quality score. This information item contains the integer product code and should be within the range 1 to 65,535.	1	-	[10, 13, 14, 15, 16, 17, 19],024, 9.135, 9..316, 14.023, 99.102- QAP-Value	ForEach(Field in [9.135, 9.316, 10.024, 16.024, 17.024, 99.102]) { ForEach(Subfield in Field) { {InfoItem:3 in Subfield} MO [1 to 65,535] AND MO [Integers] } } AND ForEach(Field in [13.024, 14.023, 14.024, 15.024, 19.024]) { ForEach(Subfield in Field) { {InfoItem:4 in Subfield} MO [1 to 65,535] AND MO [Integers] } }		
Field: Sample Quality Additional Subfield	7.7.7, 8.13.19, Table 70, 8.14.22, 8.14.23, Table 71, 8.15.17, Table73, 8.19.18 Table 85	In addition to the three information items described here, a quality field may contain other information items.	1	O	13.024- FRMP- Value	ForEach(Subfield in 13.024) {InfoItem:1 in Subfield}) MO [0 to 38, 40 to 50, 60 to 79, 81 to 84] }	B*
			1	O	14.023- FRQP-Value	ForEach(Field in [14.023]) { ForEach(Subfield in Field) {InfoItem:1 in Subfield} MO [1 to 10, 16, 17] } }	B*
			1	O	14.024- FRMP- Value	ForEach(Field in [14.024]) { ForEach(Subfield in Field) {	B*

Field	Section	Description	Count	M/O	Name	Assertion	Test	Code
						[InfoItem:1 in Subfield] MO [1 to 10, 16, 17] }}		B*
			1	O	15.024-FRMP-Value	ForEach(Subfield in 15.024) { [InfoItem:1 in Subfield] MO [20 to 38, 81 to 84] }		B*
			1	O	19.024-FRMP-Value	ForEach(Subfield in 19.024) { [InfoItem:1 in Subfield] MO [60 to 79] }		
Field: Image HLL Value	7.7.8.1	The maximum horizontal size is limited to 65,535 pixels in Record Types 4 and 8, and to 99,999 for other record types. The minimum value is 10 pixels.	1	M	[4,8].006-Value	{[4,8].006} MO [10 to 65535] AND MO [Integers]		B
			1	M	9.128-Value	{9.128} MO [10 to 99,999]		B
			1	M	xx.006-Value	{[10,13 to 17, 19, 20].006} MO [10 to 99,999] AND MO [Integers]		B
Field: Image HLL Metadata	7.7.8.1	<HLL should be checked against the image metadata to test for conformance.>		M	Fields-HLL Metadata	<The test assertions are included under field testing for the respective record types >	t-2	
Field: Image Size	7.7.8.1	The total image size (HLL times VLL) must be able to be accommodated in Field xx.001 for Traditional encoding.	3	M	xx.006,xx.007-Image Size	<Not testable due to image compression. The calculation is not simply HLL * VLL * BPX / 8>.	t-1	
Field: Image VLL Value	7.7.8.2	The maximum vertical size is limited to 65,535 pixels in Record Types 4 and 8, and to 99,999 for other record types. The minimum value is 10 pixels.	1	M	[4,8].007-Value	{[4,8].007} MO [10 to 65535] AND MO [Integers]		B
			1	M	9.129-Value	{9.129} MO [10 to 99,999]		B
			1	M	xx.007-Value	{[10,13 to 17, 19, 20].007} MO [10 to 99,999] AND MO [Integers]		B

51

Field	Clause / Table	M	Occur.	Field Code	Description	Assertion	Test	B
Field: Image VLL Metadata	7.7.8.2	M		Fields-VLL Metadata	<VLL should be checked against the image metadata to test for conformance.>	<The test assertions are included under field testing for the respective record types >	t-2	
Field: Image SLC Value	7.7.8.3, Table 27, Table 57, Table 70, Table 75, Table 85, Table 86	M	1	xx.008-Value	The image sampling frequency (pixel density).	{10,13 to 17,19,20;008} MO [0,1,2] AND MO [Integers]		B
		M	1	9.130-Value	<Tables related to each Record Type provide constraints on the value of SLC >	{9.130} MO [0,1,2] AND MO [Integers]		B
Field: Image SLC Metadata	7.7.8.3	M		Fields-SLC Metadata	A value of "1" shall indicate pixels per inch. A value of "2" shall indicate pixels per centimeter. A value of "0" in this field indicates that no scale is provided, and the quotient of THPS/TVPS shall provide the pixel aspect ratio. <SLC should be checked against the image metadata to test for conformance.>	<The test assertions are included under field testing for the respective record types >	t-2	
Field: Image SLC Contact/Contactless	7.7.8.3	M		Fields-SLC-Contactless	For contact exemplar friction ridge images, a value of 1 or 2 shall be specified…For non-contact images of body parts, SLC shall be set to 0 unless the object being imaged is a fixed distance from the capture device…	<Unsupported: Not feasible to test if the exemplar is a contact or non-contact capture.>	t-1	
Field: Image THPS Value	7.7.8.4, Table 27, Table 57, Table 70, Table 75, Table 85, Table 86	M	1	xx.009-Value	<Tables related to each Record Type provide constraints on the value of THPS.>	{10,13 to 17,19,20;009} MO [1 to 99,999] AND MO [Integers]		B
		M	1	9.131-Value		{9.131} MO [1 to 99,999] AND MO [Integers]		B
Field: Image THPS Metadata	7.7.8.4	M		Fields-THPS Metadata	This is the integer pixel density used in the horizontal direction of the image if SLC has a value of "1" or "2". If SLC has a value of "0", this information item shall contain the horizontal component of the pixel aspect ratio, up to 5 digits. <THPS should be checked against the image metadata to test for conformance.>	<The test assertions are included under field testing for the respective record types >	t-2	

Field	Reference	Description	Count	M/O	Field-Value	Assertion	Code	Type
Field: Image TVPS Value	7.7.8.5, Table 27, Table 57, Table 70, Table 75, Table 85, Table 86	If SLC is 1 or 2, then TVPS shall equal THPS. <Record layout tables provide constraints on the value of TVPS.>	1	M	xx.010-Value	{[10,13 to 17, 19, 20].010} MO [1 to 99,999] AND MO [Integers]		B
			1	M	9.132-Value	{9.132} MO [1 to 99,999] AND MO [Integers]		B
			2	M	[xx.008 to xx.010]-Conditional	IF {[10,13 to 17,19, 20].008} EQ 1 OR 2 THEN {[10,13 to 17, 19, 20].010} EQ {[10,13 to 17, 19, 20].009}		B
Field: Image TVPS Metadata	7.7.8.5	This is the integer pixel density used in the vertical direction of the image if SLC has a value of "1" or "2". If SLC has a value of "0", this information item shall contain the vertical component of the pixel aspect ratio, up to 5 integer digits. <TVPS should be checked against the image metadata to test for conformance.>		M	Fields-TVPS Metadata	<The test assertions are included under field testing for the respective record types >	t-2	
Field: Image BPX Value	7.7.8.6, 7.7.9	Some record types have a mandatory field Bits per pixel / BPX. This contains the number of bits used to represent a pixel. This field shall contain an entry of "8" for normal grayscale values of "0" to "255". Any entry in this field greater than "8" shall be used to represent a grayscale pixel with increased proportion. A maximum of 2 digits is allowed for this field. Regardless of the compression algorithm used, the image shall be represented as an array of n rows by m columns by at least 8-bit pixels. Each pixel in a grayscale image shall be represented by eight or more bits.	1	M	xx.012-Value	{[13 to 17, 19, 20].012} MO [8 to 99] AND MO [Integers]		B
Field: Image BPX Metadata	7.7.8.6	<BPX should be checked against the image metadata to test for conformance.>		M	Fields-BPX Metadata	<The test assertions are included under field testing for the respective record types >	t-2	
Field: Image SHPS Value	7.7.8.7	The horizontal pixel density used for the scanning of the original image / impression providing that the SLC field contains a "1" or "2". Otherwise, this shall indicate the horizontal component of the pixel aspect ratio, up to 5 integer digits.	1	O	xx.016-Value	{[10,13 to 16, 19, 20].016} MO [1 to 99,999] AND MO [Integers]		B
			1	O	20.017-Value	{20.017} MO [1 to 99,999] AND MO [Integers]		B

Field	Section	Description	#	M/O	Field ID	Requirement	Code	B
		This field is used if the transmission pixel scale differs from the original image scale, as listed in Transmitted horizontal pixel scale /THPS . Note that density is directly related to resolution.	1	O	17.022-Value	{17.022} } MO [1 to 99,999] AND MO [Integers]		B
Field: Image SVPS Value	7.7.8.8	The vertical pixel density used for the scanning of the original image / impression providing that the SLC field contains a "1" or "2". Otherwise, this shall indicate the vertical component of the pixel aspect ratio, up to 5 integer digits. This field is used if the transmission pixel scale differs from the original image scale, as listed in Transmitted vertical pixel scale /TVPS . Note that density is directly related to resolution. If SLC is 1 or 2 and SHPS is entered, then SVPS shall equal SHPS.	1	O	xx.017-Value	{[10,13 to 16,19].017} MO [1 to 99,999] AND MO [Integers]		B
			1	O	20.018-Value	{20.018} MO [1 to 99,999] MO [Integers]		B
			1	O	17.023	{17.023 MO [1 to 99,999] AND MO [Integers]		B
			1	O	[xx.008,xx.016, xx.017]-Conditional	IF {[10,13 to 16, 19].008} EQ 1 OR 2 THEN {[10,13 to 15].017} EQ {[10,13 to 15].016}		B
			1	O	[20.008, 20.017, 20.018]-Conditional	IF {20.008} EQ 1 OR 2 THEN {20.018} EQ {20.017}		B
			1	O	[17.008, 17.022,17.0 23]-Conditional	IF {17.008} EQ 1 OR 2 THEN {17.023} EQ {17.022}		B
Field: Image Bit Depth Value	7.7.9	Regardless of the compression algorithm used, the image shall be represented as an array of n rows by m columns by at least 8-bit pixels. Each pixel in a grayscale image shall be represented by eight or more bits.		M	Fields-Image Compressio n	<See Requirement ID "Field: Image BPX Value">	t-2	
Field: Image Format	7.7.9	Color images shall be represented as a series of sequential samples of a red, green, and blue intensity for each pixel. (Other color spaces are also possible. See Section 7.7.10.3). The image shall be organized in row-major order, with the lowest address corresponding to the upper left corner of the image. If the image is captured in grayscale, then only the luminance component shall be	3	M	Fields-Image Format	<Unsupported: The actual image data is not tested; only metadata >	t-1	

Field	Section	Requirement		Fields	Assertion / Value	Test	
Field: Image JFIF	7.7.9	compressed and transmitted. For JPEG, the data shall be formatted in accordance with the *JPEG File Interchange Format, Version 1.02 (JFIF)*.	M	Fields-Image JFIF	<This requirement applies to all instances where a JPEG image format is used. The test assertions are included under field testing for the respective record types.>	t-2	B
Field: Image Compression Algorithm Value	7.7.9.1, Table 15	For each of these fields, the entry corresponds to the appropriate *Label* entry in Table 15: -Field 13.011: Compression algorithm / CGA -Field 14.011: Compression algorithm / CGA -Field 15.011: Compression algorithm / CGA -Field 16.011: Compression algorithm / CGA (when containing a friction ridge image) -Field 19.011: Compression algorithm / CGA -Field 20.011: Compression algorithm /CGA (when containing a friction ridge image)	M	[13 to 16,19,20],0 11-Value	[[13 to 16, 19, 20],011] MO [ASCII(NONE, WSQ20, JPEGB,JPEGL, JP2, JP2L,PNG)]		1
Field: Image Compression Algorithm Metadata	7.7.9.1	<CGA should be checked against the image metadata to test for conformance.>	M	Fields-CGA Metadata	<The test assertions are included under field testing for the respective record types >	t-2	
Field: Latent Lossless Compression	7.7.9.1	Latent images shall not be compressed with any lossy compression algorithm. It is required that images be stored uncompressed, or that PNG or other totally lossless compression algorithm be used for latent images.	M	Fields-Latent Lossless Compression	<See Requirement ID's: "Field: 13.011-Compression Algorithm Value".>	t-2	
Field: Image WSQ 500ppi Exemplar	7.7.9.1	The following paragraphs apply to the exemplar images. Wavelet Scalar Quantization (WSQ) shall be used for compressing grayscale friction ridge data at 500 ppi class for new systems. In order to maintain backwards compatibility, legacy systems may use JPEGB or JPEGL for compressing 500 ppi class images. WSQ shall not be used for other than the 500 ppi class.	M	Fields-Image WSQ 500ppi Exemplar	<See Requirement ID's: "Field: 4.008-Compression Algorithm Value" and "Field: 14.011-Compression Algorithm Value".>	t-2	
Field: Image WSQ 2.0	7.7.9.1	WSQ version 3.1 or higher shall be used for WSQ compression of grayscale fingerprint data at the 500 ppi class with a	M	Fields-Image WSQ 2.0 or	<See Requirement ID's: "Field: 4.009-Image WSQ Version 2.0" and "Field: 14.999-Image WSQ Version 2.0".>	t-2, t-16	

Field								
		platen of 2 inches or greater in height. WSQ 2.0 or higher may be used for 500 ppi class data taken from a platen of less than 2 inches in height. WSQ shall not be used for other than the 500 ppi class.						
Field: Image JPEG 2000 1000ppi Exemplar	7.7.9.1	For friction ridge images at the 1000 ppi class, JPEG 2000 shall be used according to the specifications and options contained in "Profile for 1000 ppi Fingerprint Compression".	Higher	M	Fields: Image JPEG 2000 1000ppi Exemplar	<See Requirement ID: "Field: 14.011 Compression Algorithm Value".>	t-2	B
Field: Type17 Compression	7.7.9.2	For iris images, images may be uncompressed or compressed. The compression code shall be one of the following, entered in Field 17.011: Compression algorithm / CGA: - NONE – An entry of "NONE" indicates that the data contained in this record is uncompressed. - PNG – This supports lossless compression. PNG is formally standardized (ISO/IEC 15948) and implementations are freely available25 (libpng). - JP2 and JP2L - As with other biometrics, while lossless compression is preferred, iris images can be lossy-compressed. The image type (Field 17.032: Iris storage format / ISF) should be selected appropriately, and the compression ratio should be set to satisfy some known quantified storage or transmission bandwidth limitation. The baseline JPEG algorithm (ISO/IEC 10918) is not acceptable for iris images and shall not be used. It has been shown that false match rates increase due to the presence of tiling artifacts. While JPEG was allowed in prior versions of this standard for iris compression, it is not allowed for this version. Implementers may want to support JPEG decoding for handling legacy images.	1	M	17.011-Value	{17.011} MO [ASCII(NONE, JP2,JP2L,PNG)]		
Field: Type10 Facial Compression	7.7.9.3, E.6.1	When Record Type-10 contains facial image, the conditions described in Annex E: E.6.1 Compression algorithm apply.	3	M	Fields: Type 10 Facial Compression	<Unsupported: Not feasible to determine if the images are non-frontal or conform to the SAP level.>	t-9	

56

Field	Section	Description	Count	M	Value	Assertion	
Field: Type10 Compression n	7.7.9.4	For non-facial images contained in Record Type-10, Field 10.011: Compression algorithm / CGA may be set to any value in Table 15, except WSQ20.	1	M	10.011-Value	IF {10.003} NEQ ASCII(FACE) THEN {10.011} MO [ASCII(NONE, JPEGB, JPEGL, JP2, JP2L, PNG)] AND IF {10.003} EQ ASCII(FACE) THEN <Depends on SAP Level, Currently Not Supported-Provide Warning>	B
Field: Type16 Compression n	7.7.9.4	Non-friction ridge images contained in Record Type-16 shall specify the file extension (suffix) corresponding to the compression used, such as OOG, JPG, WAV, and PNG in Field 16.011: Compression algorithm / CGA. A value of "NONE" indicates that the data is uncompressed.	1	M	16.011-Value	<The test assertions for this type may not be supported in this version of the CTM. If they are supported, they are included under field testing for Record Type-16: User-defined testing image record. >	t-2
Field: Image CSP Value	7.7.10.3, Table 16	Several image record types have a field Color space / CSP. It shall contain an entry from the CODE column of Table 16. If the color image type cannot be determined, an entry of "RGB" shall be entered in this field.	1	M	10.012-Value	{10.012} MO [ASCII(UNK, GRAY, RGB, SRGB, YCC, SYCC)]	B
			1	M	xx.013-Value	{[16,17,20].013} MO [ASCII(UNK, GRAY, RGB, SRGB, YCC, SYCC)]	B
Field: Image ECL Value	7.7.11, Table 17	This information appears in Field 10.027: Subject eye color / SEC and in Field 17.020: Eye color / ECL. The eye color describes the eye color of the subject as seen in the image. If unusual or unnatural, such as is the case when colored contact lenses are present and the 'real' eye color cannot be ascertained, then the color shall be labeled as "XXX". For near infra-red (NIR) images, if this field is entered, it shall be "XXX'. Values for these fields shall be the alphabetic entries in the "Attribute code" column of Table 17.	1	-	10.027,17.020-Value	{10.027} AND {17.020} MO [ASCII(BLK, BLU, BRO, GRY, GRN, HAZ, MAR, MUL, PNK, XXX)]	B
Field: Image-Paths	7.7.12	A path may not have any sides crossing. No two vertices may occupy the same position. There may be up to 99 vertices. An open path is a series of connected line segments that do not close or overlap. A	-		Fields-Image-Paths	<The test assertions are included under field testing for the respective record types >	t-2

Field: Image-EFS Paths	7.7.12.1 Table 30	-	Fields-Image-EFS Paths	t-2
closed path (polygon) completes a circuit. The closed path side defined by the last vertex and the first vertex shall complete the polygon. A polygon shall have at least 3 vertices. The contours in Record Type-17: Iris image record can be a circle or ellipse. A circle only requires 2 points to define it (See Table 19). There are two different approaches to the paths in this standard. The 2007 and 2008 version of the standard used paths for Field 14.025: Alternate finger segment position(s) / ASEG. That approach has been retained in this version for all paths except in the Extended Feature Set (EFS) of Record Type-9.	The vertices for paths in the EFS Type-9 records are defined in a single information item27 for each of the following fields (See Table 30 Type-9 Fields for EFS). If multiple paths are present, they are stored within separate subfields. Each vertex is expressed as an (X,Y) pair of positive integers in units of 10 micrometers (0.01mm). The Extended Feature Set used in the Record Type-9: Minutiae data record was developed as a separate encoding structure that has been incorporated into this standard. In order to avoid conflicts with systems that had already programmed using the EFS method of specifying paths, that structure is retained in this standard. EFS fields using closed paths, and requiring at least 3 vertices, are: • Field 9.300: EFS region of interest / ROI • Field 9.302: EFS finger – palm – plantar position / FPP • Field 9.360: EFS area of correspondence / AOC An open path is a series of • Field 9.324: EFS distinctive features / DIS • Field 9.357: EFS local quality issues / LQI An open path is a series of connected		<The test assertions are included under field testing for the respective record types >	

| Field: Image- Other Paths | 7.7.12.2, Tables 18 to 21 | points in which there is not an implicit connection between the last and first vertices. Within EFS, open paths are used in Field 9.373: EFS ridge path segments / RPS. The first information item is dependent upon the Record Type and field. The common format is prefixed by no, one or two information items, depending upon the field. | - | Fields-Image-Other Paths | <The test assertions are included under field testing for the respective record types > | t-2 | | | | |

Table 6.4 - Assertions for Record Type 1- Transaction Information Record

Requirement ID	Reference in Base Standard	Requirement Summary	Assertion ID	Level	Status	Test Assertion	Test Note	Implementation Support	Supported Range	Test Result	Applicability
						8.1: Record Type-1: Transaction information record					
Transaction: Type1 Mandatory	8.1	Record Type-1 is mandatory. Only one Type-1 record is present per transaction.	Type1-Mandatory		M	<See Requirement ID "Transaction: Type1-Occurrences">	t-2				
Record: Type1 7-bit ASCII	8.1	Note that since the alternate character encoding is specified in this record, t here must be specified characters agreed upon in order to read this Record Type, particularly with Traditional encoding, and the characters that can be represented by the 7-bit ASCII code are those characters (see Table 93 for these characters).	Type1-ASCII		M	<The assertions check for the more specific character types allowed by Table 22. See Requirement IDs: : "Field: Type1-Char Type" and "Field: Type1-Subfield Char Type" >	t-2				
Field: Type1-Subfield Occurrence	Table 22	<Table 22 specifies which fields contain subfields as well as the number of occurrences permitted >	1.[001, 002, 004 to 015, 017]-SubfieldCount	1	M	Count(Subfields in 1.[001, 002, 004 to 015, 017]) EQ 1					B*
			1.[001, 002, 004 to 012]-InfoItemCount	1	M	Count(InfoItems in Subfield:1 in 1.[001, 002, 004 to 012, 014]) EQ1					B*
			1.003-	1	M	ForEach(Subfield in 1.003)					T

Field	Reference	Description	Element	Condition		#	Type
			InfoItemCount	{ Count(InfoItems in Subfield) EQ 2 }			
			1.003-SubfieldCount	Count(Subfields in 1.003) GTE 2	M	1	B*
			1.013-InfoItemCount	Count(InfoItems in 1.013) EQ 1 OR 2	O	1	B*
			1.015-InfoItemCount	Count(InfoItems in 1.015) MO [2, 3]	O	1	B*
			1.015-SubfieldCount	Count(Subfields in 1.015) EQ 1	O	1	B*
			1.016-InfoItemCount	Count(InfoItems in 1.016) EQ 3 * Count(Subfields in 1.016)	O	1	B*
			1.016-Subfield Count	Count(Subfields in 1.016) MO [1 to 99]	O	1	B*
			1.017-InfoItemCount	Count(InfoItems in 1.017) LTE 2	O	1	T
Field: Type1-CondCode	Table 22	<Table 22 specifies the Condition Code for each field.>	1.[001 to 005, 007 to 009, 011, 012]-Mandatory CondCode	Present(1.001 to 1.005, 1.007 to 1.009, 1.011, 1.012)	M	1	B
Field: Type1-CharType	8.1, Table 22	<Table 22 specifies the Character Type for each field that contains no subfields.>	1.[001, 002, 006]-CharType	Bytes(1.[001, 002, 006]) MO [0x30 to 0x39]	-	1	B
			1.004-CharType	Bytes(1.004) [0x20, 0x41 to 0x5A, 0x61 to 0x7A]	-	1	B
			1.005-CharType	Bytes(1.005) MO [0x30 to 0x39]	M	1	T
			NIEM-1.005-CharType	Bytes(1.005) MO [0x30 to 0x39, 0x2D]	M	1	X
			1.007, 1.008-CharType	Bytes(1.007, 1.008) MO [0x20 to 7E]	M	1	B
			1.009, 1.010-CharType	Bytes(1.[009,010]) MO [0x30 to 0x39, 0x20, 0x41 to 0x5A, 0x61 to 0x7A]	-	1	B

Field	Reference	Description			Element	Condition	Type
			1	-	1.011, 1.012-CharType	Bytes(1.[011,012]) MO [0x30 to 0x39, 0x2E]	B
			1	O	1.014-CharType	Bytes(1.014) MO [0x30 to 0x39,0x5A]	B
Field: Type1-Subfield CharType	8.1, Table 73	<Table 22 specifies the Character Type for each subfield.>	1	O	1.003-[FRC, CRC, REC, IDC]-CharType	Bytes(All(Subfields in 1.003)) MO [0x30 to 0x39]	B*
			1	O	1.013-[DNM, DVM]-CharType	ForEach(InfoItem in 1.013) { Bytes(InfoItem) MO [0x20 to 0x7E] }	B*
			1	-	1.015-CSI-CharType	Bytes(InfoItem:1 in 1.015) MO [0x30 to 0x39]	B*
			1	-	1.015-[CSN, CSV]-CharType	Bytes(InfoItem:2,3 in 1.015) MO [0x20 to 0x7E]	B*
			1	-	1.016-[APO, APN, APV]-CharType	Bytes(All(Subfields in 1.016)) MO [0x20 to 0x7E]	B*
			1	-	1.017-[DAN, OAN]-CharType	Bytes(All(Subfields in 1.017)) MO [0x20 to 0x7E]	B*
Field: Type1-CharCount	Table 22	<Table 22 specifies the Character Count for each field that contains no subfields.>	1	M	1.001-CharCount	DataLength(1.001) GTE 2	T
			1	M	NIEM-1.001-CharCount	Length(1.001) EQ 2	X
			1	M	1.002-CharCount	DataLength(1.002) EQ 4	T
			1	M	NIEM-1.002-CharCount	Length(1.002) EQ 3	X
			1	M	1.004-CharCount	DataLength(1.004) GTE 1 AND LTE 16	B
			1	M	1.005-CharCount	DataLength(1.005) EQ 8	T
			1	M	NIEM-1.005-CharCount	DataLength(1.005) EQ 10	X
			1	O	1.006-CharCount	DataLength(1.006) EQ 1	B
			1	O	1.007-	DataLength(1.007) GTE 1	B

Field		Table 22						Type
						CharCount		
			1	O	1.008-CharCount	DataLength(1.008) GTE 1		B
			1	M	1.009-CharCount	DataLength(1.009) GTE 1		B
			1	O	1.010-CharCount	DataLength(1.010) GTE 1		B
			1	M	1.011-CharCount	DataLength(1.011) EQ 5		T
			1	M	NIEM-1.011-CharCount	Length(1.011) EQ 4		X
			1	M	1.012-CharCount	DataLength(1.012) EQ 5		T
			1	M	NIEM-1.012-CharCount	Length(1.012) EQ 4		X
			1	O	1.014-CharCount	DataLength(1.014) EQ 15		T
			1	O	NIEM-1.014-CharCount	DataLength(1.014) EQ 20		X
Field: Type-1 Subfield CharCount	Table 22	<Table 22 specifies the Character Count for each subfield.>	1	M	1.003-FRC-CharCount	Length(InfoItem:1 in Subfield:1 in 1.003) EQ 1		B*
			1	M	1.003-CRC-CharCount	Length(InfoItem:2 in Subfield:1 in 1.003) GTE 1 AND LTE 3		B*
			1	M	1.003-[REC,IDC]-CharCount	ForEach(Subfield in 1.003 ST Subfield NOT Subfield:1 in 1.003) { Length(InfoItem:1,2 in Subfield) GTE 1 AND LTE 2 }		B*
			1	O	1.013-[DNM,DVN]-CharCount	Length(InfoItems in 1.013) GTE 1		B*
			1	O	1.015-CSI-CharCount	Length(InfoItem:1 in 1.015) GTE 1 AND LTE 3		B*
			1	O	1.015-[CSN,CSV]-CharCount	Length(InfoItem:2,3 in 1.015) GTE 1 AND LTE 16		B*
			1	O	1.016-[APO, APN, APV]-CharCount	ForEach(Subfield in 1.016) { Length(InfoItems in Subfield) GTE 1 }		B*

Field	Reference	Description	Occ	M/O	Field	Condition / Length		B*
			1	O	1.017-[DAN,OAN]-CharCount	Length(InfoItems in 1.017) GTE 1		B*
Field: Type1-Field Occurrence	Table 22	<Table 22 specifies the Field Occurrence for each field.>						B
			1	M	1.001-Occurrence	Count(1.001) EQ 1		B
			1	M	1.002-Occurrence	Count(1.002) EQ 1		B
			1	M	1.003-Occurrence	Count(1.003) EQ 1		B
			1	M	1.004-Occurrence	Count(1.004) EQ 1		B
			1	M	1.005-Occurrence	Count(1.005) EQ 1		B
			1	O	1.006-Occurrence	Count(1.006) LTE 1		B
			1	M	1.007-Occurrence	Count(1.007) EQ 1		B
			1	M	1.008-Occurrence	Count(1.008) EQ 1		B
			1	M	1.009-Occurrence	Count(1.009) EQ 1		B
			1	O	1.010-Occurrence	Count(1.010) LTE 1		B
			1	M	1.011-Occurrence	Count(1.011) EQ 1		B
			1	M	1.012-Occurrence	Count(1.012) EQ 1		B
			1	O	1.013-Occurrence	Count(1.013) LTE 1		B
			1	O	1.014-Occurrence	Count(1.014) LTE 1		B
			1	O	1.015-Occurrence	Count(1.015) LTE 1		B
			1	O	1.016-Occurrence	Count(1.016) LTE 1		B
			1	O	1.017-Occurrence	Count(1.016) LTE 1		B
Field: 1.001-Record Header Value	8.1.1, Table 22, 7.1	Field 1.001 Record header. In Traditional encoding, this field contains the record length in bytes (including all information separators)	2	M	1.001-Record Header	<See Requirement ID "Field: xx.001-Record Header">	t-2	
	8.1.1, C.9.1	The XML name for the Type-1 record is <itl:PackageInformationRecord>, and its <biom:RecordCategoryCode> element shall have a value of 01.	1	M	NIEM-1.001-Value	ForEach(XElm(itl:PackageInformationRecord) { (XElm(biom:RecordCategoryCode) EQ ASCII(01) }		X

Field	Reference	#	Status	Requirement ID	Description	Test		Type
Field: 1.002-Version Number Value	8.1.2, Table 22	1	M	1.002-Value	This mandatory four-character ASCII value shall be used to specify the current version number of the standard implemented by the software or system creating the transaction. The format of this field shall consist of four numeric characters. The first two characters shall specify the major version number. The last two characters shall be used to specify the minor revision number. This version of the standard has the entry "0500"	{1.002} EQ ASCII(0500)		B
Field: 1.003-Transaction Content Subfields	8.1.3, Table 22	1	M	1.003-SubfieldCount	This mandatory field shall list and identify each of the records in the transaction by record type and its IDC value. It also specifies the order in which the remaining records shall appear in the file. It shall consist of two or more subfields. The first subfield shall relate to this Type-1 Transaction record.	<See Requirement ID "Field: Type 1-Subfield Occurrence" and "Field: xx.002-IDC">	t-2	
Field: 1.003-Transaction Content Subfield 1 FRC Value	8.1.3, Table 22	1	M	1.003-FRC-Value	The first information item (record category code / REC) within this subfield shall be "1". This indicates that the first record in the transaction is a Type-1 record consisting of header information.	{InfoItem:1 in Subfield:1 in 1.003} EQ 1		B*
Field: 1.003-Transaction Content Subfield 1 CRC Value	8.1.3, Table 22	1	M	1.003-CRC-Value	The second information item of this subfield (content record count / CRC) shall be the sum of the Type-2 through Type-99 records contained in this transaction. This number is also equal to the count of the remaining subfields of Field 1.003 Transaction content / CNT. The maximum value for CRC is 999.	{InfoItem:2 in 1.003} MO [Integers] AND MO [1 to 999]		B*
		2	M	1.003-CRC-Transaction Record Count		{InfoItem:2 in 1.003} EQ Count(Records in Transaction ST Type(Records) MO [2 to 99])		T
		2	M	1.003-CRC-Matches Subfields		{InfoItem:2 in 1.003} EQ Count(Subfields in 1.003) EQ - 1		T
Field: 1.003-Transaction Content Subfield 2 REC Value	8.1.3, Table 22, Table 3	1	M	1.003-REC-Value	Each of the remaining subfields of Field 1.003 Transaction content / CNT corresponds to a single Type-2 through Type-99 record contained in the transaction. Two information items shall comprise each of these subfields: The first information item (record category code / REC), shall contain a number chosen from the "record"	ForEach(Subfield in 1.003 ST Subfield NOT Subfield:1 in 1.003) { {InfoItem:1 in Subfield} MO [2,4,7 to 10, 13 to 21, 98,99] AND MO [Integers] }		B*
		1	M	1.003-REC-Matches Records		ForEach(Subfield in 1.003 ST Subfield NOT Subfield:1 in 1.003) {		B*

Field	Reference	Description	#	M/O	Value	Condition	Type	
		"identifier" column of Table 3. It also specifies the order in which the remaining records shall appear in the file.				Present(Record in Transaction ST Type(Record) EQ {InfoItem:1 in Subfield}} <The records and associated record numbers must be in the same order >		

Field: 1.003-Transaction Content Subfield 2 IDC Value	8.1.3, Table 22	The second information item (information designation character / IDC) shall be an integer equal to or greater than zero and less than or equal to 99. See Section 7.3.1.	1	M	1.003-IDC-Value	ForEach(Subfield in 1.003 ST NOT Subfield:1 in 1.003)
						 {
						 {InfoItem:2 in Subfield} MO [0 to 99] AND MO [Integers]
						 }
							B*	

Field: 1.003-Transaction Content Subfield 2 IDC Matches	8.1.3, Table 22, 7.3.1	IDC references are stated in Type-1 Field 1.003 Transaction content / CNT and shall be used to relate information items in the CNT field of the Type-1 record to the other records in the transaction. It also specifies the order in which the remaining records shall appear in the file.	2	M	1.003-IDC-Matches Records	ForEach (Record in Transaction) { Present{Subfield in 1.003 ST Subfield NOT Subfield:1 in 1.003 AND {InfoItem:1 in Subfield} EQ Type(Record) AND {InfoItem:2 in Subfield} EQ {Record.002} <The records and associated record numbers must be in the same order >	B*	
Field: 1.004-Type of Transaction Value	8.1.4, Table 22	This mandatory field shall contain an identifier, which designates the type of transaction and subsequent processing that this transaction should be given. This shall be a maximum of 16 alphabetic characters. The TOT shall be in accordance with definitions provided by the receiving agency.) Earlier versions of this standard specifically restricted the character length of TOT to 4 characters.	1	M	1.004-Value	TRUE	B	
Field: 1.005-Local Date Value	8.1.5, Table 22, 7.7.2.3	This mandatory field shall contain the local date that the transaction was submitted. The local date is recorded as YYYYMMDD. Note that this may be a different date than the corresponding GMT, due to time zone differences.	1	M	1.005-Value	{1.005} MO [ValidLocalDate]	T	t-6
			1	M	NIEM-1.005-Value	ForEach(XElm(itl::PackageInformationRecord)) {XElm(nc:Date) in XElm(biom:TransactionDate)} MO [NIEM-ValidLocalDate] }	X	t-6
Field: 1.006-Priority Value	8.1.6, Table 22	This optional field shall contain a single information character to designate the urgency with which a response is desired. The values shall range from 1 to 9, with 1 denoting the highest priority. The default	1	O	1.006-Value	{1.006} MO [1 to 9] AND MO [Integers]	B	

Field	Reference	Description	Count	M/O	Field-Value			Requirement	Code
		value shall be defined by the agency receiving the transaction.							
Field: 1.007-Destination Agency Identifier Value	8.1.7, Table 22	This mandatory field shall contain the identifier of the administration or organization designated to receive the transmission. The size and data content of this field shall be user-defined and in accordance with the application profile.		M	1.007-Value	TRUE			B
Field: 1.008-Originating Agency Identifier Value	8.1.8, Table 22	This mandatory field shall contain the identifier of the administration or organization originating the transaction. The size and data content of this field shall be user-defined and in accordance with the application profile.		M	1.008-Value	TRUE			B
Field: 1.009-Transaction Control Number Value	8.1.9, Table 22	This mandatory field shall contain the transaction control number as assigned by the originating agency. A unique (for the originating agency) alphanumeric control number shall be assigned to each transaction. For any transaction that requires a response, the respondent shall refer to this number in communicating with the originating agency.	1	M	1.009-Value	TRUE			B
Field: 1.010-Transaction Control Reference Value	8.1.10, Table 22	This optional field shall be used for responses that refer to the TCN of a previous transaction involving an inquiry or other action that required a response.	1	O	1.010-Value	TRUE			B
Field: 1.011-Native Scanning Resolution Value	8.1.11, Table 22	This mandatory field shall be set to "00.00" if there are no Type-4 records in the transaction. When there are Type-4 records present, this field is used to specify the native scanning resolution of the friction ridge image capture device. This field shall specify the resolution in pixels per millimeter. The resolution shall be expressed as two numeric characters followed by a decimal point and two more numeric characters.		M	1.011-NSR	<See Requirement IDs "Field: NSR Conditional".>				t-2
Field: 1.012-Nominal Resolution Value	8.1.12, Table 22	This mandatory field shall be set to "00.00" if there are no Type-4 records in the transaction. When there are Type-4 records present, this field specifies the nominal resolution for the image(s) being exchanged. This field shall specify the resolution in pixels		M	1.012-NTR	<See Requirement IDs "Field: Type4 NTR" and "Record: Type4-Resolution 500 Only"..>				t-2

Field	Reference	#	Char	Field ID	Description	Value / Condition	t	Type
					per millimeter. The resolution shall be within the range 19.30 ppmm (490 ppi) to 20.08 ppmm (510 ppi).			B*
Field: 1.013-Domain Name Value	8.1.13, Table 22	1	O	1.013-DOM	The mandatory first information item (domain name / DNM) will uniquely identify the agency, entity, or implementation used for formatting the fields in the Type-2 record. The default value for the field shall be the North American Domain implementation (NORAM). An optional second information item (domain version number / DVN) shall contain the unique version of the particular implementation, such as 7.02.	TRUE		
Field: 1.013-Domain Name Occurs Once	8.1.13,		O	1.013-DOM-Occurs Once	The domain name may only appear once within a transaction.	<See Requirement ID "Field: Type 1-Field Occurrence">	t-2	
Field: 1.014-Greenwich Mean Time Value	8.1.14, Table 22	1	O	1.014-GMT	This optional field provides a mechanism for expressing the date and time in terms of universal Greenwich Mean Time (GMT) units.	{1.014} MO [ValidUTC/GMT]	t-6	T
		1	O	NIEM-1.014-GMT		ForEach(XElm(it:PackageInformationRecord)) { XElm(nc:DateTime) in XElm(biom:TransactionUTCDate)} MO [NIEM-ValidUTC/GMT] }	t-6	X
Field: 1.015-Character Encoding Subfield 1 CSI Value	8.1.15, Table 22, Table 4	1	M ⇑	1.015-DCS-CSI-Value	The first information item (character encoding index / CSI) is the index number that references an associated character encoding. See the "Character encoding index" column of Table 4 for the valid values for this information item.	{InfoItem:1 in 1.015} MO [0 to 4, 128 to 999] AND MO [Integers] <If the value '1' is present, display a warning that the value is a Legacy value only.>		B*
Field: 1.015-Character Encoding Subfield 2 CSN Value	8.1.15, Table 22, Table 4	2	M	1.015-DCS-CSN-Value	The second information item (character encoding name / CSN) shall be the "Character encoding name" associated with that index number, taken from Table 4.	IF {InfoItem:1 in 1.015} EQ 0 THEN {InfoItem:2 in 1.015} EQ ASCII(ASCII) AND IF {InfoItem:1 in 1.015} EQ 1 THEN {InfoItem:2 in 1.015} EQ ASCII(8-bit ASCII) AND IF {InfoItem:1 in 1.015} EQ 2 THEN		B*

Field	Reference	Count		Mnemonic	Value / Condition		Code	Note
Field: 1.015-Character Encoding Subfield 3 CSV Value	8.1.15, Table 22, Table 4	1	O ⇑	1.015-DCS-CSV-Value	The optional third information item (character encoding version / CSV) is the specific version of the character encoding used. In the case of the use of UTF-8, the third optional information item may be used to hold the specific version used, so that the display terminal can be switched to the correct font family.	TRUE	t-1, t-4	B* {InfoItem:2 in 1.015} EQ ASCII(UTF-16) AND IF {InfoItem:1 in 1.015}EQ 3 THEN {InfoItem:2 in 1.015} EQ ASCII(UTF-8) AND IF {InfoItem:1 in 1.015} EQ 4 THEN {InfoItem:2 in 1.015} EQ ASCII(UTF-32)
Field: 1.015-Character Encoding User Defined	8.1.16, Table 22	3	O	1.015-DCS-User Defined Encoding	This optional field specifies the character encoding that may appear within this transaction for data with the character type listed as "U" or 'user-defined' in the record format tables.	<Unsupported>		
Field: 1.016-Application Profile Specifications Value	8.1.16, Table 22	1	O	1.016-APS	There may be multiple subfields, each designating an application profile to which this transaction conforms Each subfield shall consist of three mandatory information items: The first information item (application profile organization / APO) will uniquely identify the agency or entity responsible for the specification. The second information item (application profile name / APN) shall contain the name of the specification. The third information item (application profile version number / APV) shall contain the specific version of the specification.	TRUE	t-1, t-3	B*
Field: 1.016-Application Profile	8.1.16	3	O	1.016-APS Compliance	If multiple Application Profile Specifications are included in this field, the specifications must be compatible	<Unsupported>		

Specifications Compliance		with each other: this transaction must be in compliance with all of the cited specifications. See Section 6.					B
Field:1.017-Agency Names Value	8.1.17	Both information items are alphanumeric and can have any special characters in the names.	1	O	1.017-ANM	TRUE	

Table 6.5 - Assertions for Record Type 3 – Deprecated

Requirement ID	Reference in Base Standard	Requirement Summary	Lvl	Status	Assertion ID	Test Assertion	Implementation Support	Supported Range	Test Note	Test Result	Applicability
						8.3: Record Type-3: DEPRECATED					
Transaction: Type3 Zero Occurrences	8.3	No instances of Record Type-3 shall be included in a transaction conformant with this version of the standard.	2	M	Type3-Zero Occurrences	ForEach(Record in Transaction) { Type(Record) NEQ 3 }					T
			2	M	NIEM-Type3-Zero Occurrences	<An invalid record type will cause a parsing error in XML because no Record Element Tag is defined (see Table 100).>					X

Table 6.6 - Assertions for Record Type 4 - Grayscale Fingerprint Image

Requirement ID	Reference in Base Standard	Requirement Summary	Lvl	Status	Assertion ID	Test Assertion	Implementation Support	Supported Range	Test Note	Test Result	Applicability
						8.4: Record Type-4: Grayscale fingerprint image					
Record: Type4 Scan Resolution 500ppi	8.4	The Type-4 record is based on the use of a captured fingerprint image obtained using a class scanning resolution of the 500 ppi class. (See Section 7.7.6). It shall not be used for other than 500 ppi. class images.		M	Type4-Scan Resolution 500ppi	<See Requirement ID "Field: Type4NTR">			t-2		
Record: Type4 Image Compression WSQ	8.4	All images that are compressed shall be compressed using WSQ. JPEG compression is retained solely for backwards compatibility with legacy systems and it should not be used in any		M	Type4-CGA WSQ Only	<See Requirement IDs "Field: 4.008-Compression Algorithm Value" and "Field: 4.009-Image Data Valid">.			t-2		

Field	Reference	Description	Count	M	Field ID	Rule	Code	Type
		new implementation.						
Field: Type4-CondCode	Table 24, Table 98	<Table 24 specifies the Condition Code for each field.>	1	M	4.[001 to 009]-Mandatory CondCode	Present(4.001 to 4.009)		B
Field: Type4-CharType	Table 24, Table 98	Traditional format requires the data in binary form (not text) with a fixed byte length.	3	M	Type4-CharType	<Not tested directly, but Type 4 is parsed as untagged binary data for Traditional encoding>.	t-13	T
		<Table 24 specifies the Character Type for each field.>	1	M	NIEM-Type4-CharType	Bytes(4.[001 to 008]) MO [0x30 to 0x39]		X
Field: Type4-ByteCount	Table 24, Table 98	<Table 98 specifies the Byte Count for each field.>	1	M	4.001-ByteCount	Length(4.001) EQ 4		T
			1	M	NIEM-4.001-ByteCount	ForEach(XElm(itl:PackageHighResolutionGrayscaleImageRecord)) { Length(XElm(biom:RecordCategoryCode) EQ 2 }		X
			1	M	4.002-ByteCount	Length(4.002) EQ 1		T
			1	M	NIEM-4.002-ByteCount	ForEach(XElm(itl:PackageHighResolutionGrayscaleImageRecord)) { Length(XElm(nc:IdentificationID) in XElm(biom:ImageReferenceIdentification)) EQ 1 OR 2 }		X
			1	M	4.003-ByteCount	Length(4.003) EQ 1		T
			1	M	NIEM-4.003-ByteCount	ForEach(XElm(itl:PackageHighResolutionGrayscaleImageRecord)) { Length(XElm(biom:FingerprintImageImpressionCaptureCategoryCode)) EQ 1 OR 2 }		X
			1	M	4.004-ByteCount	Length(4.004) EQ 1		T
			1	M	NIEM-4.004-ByteCount	ForEach(XElm(itl:PackageHighResolutionGrayscaleImageRecord)) { Length(All(XElm(biom:FingerPositionCode) in XElm(biom:FingerprintImagePosition))) MO [1 to 3] }		X
			1	M	4.005-ByteCount	Length(4.005) EQ 1		T

1	M	NIEM-4.005-ByteCount	ForEach(XEIm(itl:PackageHighResolutionGrayscaleImageRecord)){ Length(XEIm(biom:CaptureResolutionCode) in XEIm(biom:ImageCaptureDetail)) EQ 1 }	X
1	M	4.006-ByteCount	Length(4.006) EQ 2	T
1	M	NIEM-4.006-ByteCount	ForEach(XEIm(itl:PackageHighResolutionGrayscaleImageRecord)){ Length(XEIm(biom:ImageHorizontalLineLengthPixelQuantity)) MO [2 to 5] }	X
1	M	4.007-ByteCount	Length(4.007) EQ 2	T
1	M	NIEM-4.007-ByteCount	ForEach(XEIm(itl:PackageHighResolutionGrayscaleImageRecord)){ Length(XEIm(biom:ImageVerticalLineLengthPixelQuantity)) MO [2 to 5] }	X
1	M	4.008-ByteCount	Length(4.008) EQ 1	T
1	M	NIEM-4.008-ByteCount	ForEach(XEIm(itl:PackageHighResolutionGrayscaleImageRecord)){ Length(XEIm(biom:ImageCompressionAlgorithmCode)) EQ 1 }	X
2	M	4.009-ByteCount	Length(4.009) EQ (4.001)−18	T
1	M	NIEM-4.009-ByteCount	ForEach(XEIm(itl:PackageHighResolutionGrayscaleImageRecord)){ Length(XEIm(nc:BinaryBase64Object)) GTE 1 }	X
1	M	4.[001 to 003, 005 to 009]-Occurrence	Count(4.[001 to 003, 005 to 009]) EQ 1	B
1	M	4.004-Occurrence	Count(4.004) EQ 6	T
1	M	4.004-Occurrence NIEM	Count(4.004) MO [1 to 6] ForEach(XEIm(itl:PackageHighResolutionGrayscal	X

Field: Type 4-Field Occurrence

Table 24 <Table 24 specifies the Field Occurrence for each field.>

Field	Reference	Description		M	Field ID	Content		Type
						elmageRecord) { Count(XElm(biom:FingerPositionCode)) MO [1 to 6] }		
Field: 4.001-Record Header Value	8.4.1, Table 24, 7.1	Field 4.001 Record header. In Traditional encoding, this field contains the record length in bytes (including all information separators)		M	4.001-Record Header	<See Requirement ID "Field: xx.001-Record Header">	t-2	X
	8.4.1, C.9.4	The XML name for the Type-4 record is <itl:PackageHighResolutionGrayscaleRecord>, and its <biom:RecordCategoryCode> element shall have a value of "04".	1	M	NIEM-4.001-Value	ForEach(XElm(itl:PackageHighResolutionGrayscaleRecord) { {XElm(biom:RecordCategoryCode)} EQ ASCII(04) }		
	Table 24	<The value in Field 4.001 must be 19 or greater due to the character count of fields in the Type 4 Record>	1	M	4.001-Value	{4.001} GTE 19		T
Field: 4.002-Information Designation Character Value	8.4.2, Table 24, 7.3.1	This mandatory field shall contain the IDC assigned to this Type-4 record as listed in the information item IDC for this record in Field 1.003 Transaction content/CNT.	1	M	4.002-Value	<See Requirement IDs "Field: xx.002-IDC and "Field: 1.003-Transaction Content Subfield 2 IDC Matches">	t-2	
Field: 4.003-Impression Type Value	8.4.3, Table 24, 7.7.4.1	This mandatory field shall indicate the manner by which the fingerprint was obtained. See Section 7.7.4.1 for details.	1	M	4.003-Value	{4.003} MO [0 to 3, 8, 20 to 29] AND MO [Integers]		B
Field: 4.004-Friction Ridge Generalized Position Value	8.4.4, Table 24, Table 8, 7.7.4.2	This mandatory field shall contain the decimal code number corresponding to the finger position and shall be taken from Table 8 (only finger numbers 0-15 apply to Type-4). Up to five additional finger positions shall be referenced by entering the alternate finger positions using the same format. If fewer than five finger position references are to be used, the unused position references shall be filled with 255. Six values shall be entered in each record.	1	M	4.004-Value	{4.004} MO [0 to 15, 255]		T
			1	M	NIEM-4.004-FGP	ForEach(XElm(itl:PackageHighResolutionGrayscaleRecord) {All(XElm(biom:FingerPositionCode) in XElm(biom:FingerprintImagePosition))} MO [0 to 15, 255] AND MO [Integers] }		X
Field: 4.005-Image Scanning Resolution Value	8.4.5, Table 24	The mandatory ISR field relates to the scanning resolution of this image.	1	M	4.005-Value	{4.005} MO [0,1]		B
		Previous versions of this standard stated that 0 in this field represents the 'minimum scanning resolution.' The minimum scanning resolution was defined in ANSI/NIST-ITL 1-2007 as "19.69 ppmm plus or minus 0.20 ppmm (500 ppi	2	M	4.005-Conditional	IF {1.011} LT 19.49 OR GT 19.89 THEN {4.005} EQ 1 ELSE {4.005} EQ 0		B

plus or minus 5 ppi)." Therefore, if the image scanning resolution corresponds to the Appendix F certification level (See Table 14* Class resolution with defined tolerance), a 0 shall be entered in this field.
A value of 1 is entered if the actual scanning resolution (outside of the Appendix F certification range) is specified in Field 1.011 Native scanning resolution / NSR.

Field	Ref	Description		Req ID	Requirement	Code	
Field: 4.006- Horizontal Line Length Value	8.4.6, Table 24	This mandatory field shall contain the number of pixels on a single horizontal line of the transmitted image.	M	4.006-Value	<See Requirement ID "Field: Image HLL Value">	t-2	
Field: 4.006- Horizontal Line Length Metadata	8.4.6, Table 24, WSQ Standard	<The HLL is verified by checking the image metadata if compression is used. >	2	4.006-Matches Metadata	IF {4.008} EQ 1 THEN {4.006} EQ {Image Width-WSQ} ELSE IF {4.008} EQ 2 OR 3 THEN {4.006} EQ {Image Width-JPEGB, JPEGL}	t-11	B
Field: 4.007- Vertical Line Length Value	8.4.7, Table 24	This mandatory field shall contain the number of pixels on a single horizontal line of the transmitted image.	M	4.007-VLL	<See Requirement ID "Field: Image VLL Value">		
Field: 4.007- Vertical Line Length Metadata	8.4.7, Table 24	<The VLL is verified by checking the image metadata if compression is used.>	2	4.007-Matches Metadata	IF {4.008} EQ 1 THEN {4.007} EQ {Image Height-WSQ} ELSE IF {4.008} EQ 2 OR 3 THEN {4.007} EQ {Image Height-JPEGB, JPEGL}	t-11	B
Field: 4.008- Compression Algorithm Value	8.4, 8.4.8, Table 24	All images shall be compressed using WSQ. This is a mandatory field, used to specify the type of compression algorithm used. A zero denotes no compression. Otherwise, the WSQ algorithm should be used to compress the data, and is indicated by a value of 1. Codes 2 and 3 are retained solely for backwards compatibility with those legacy systems that use JPEG compression and should not normally be used.	1	4.008-Value	IF {4.008} EQ 2 OR 3 <Provide Legacy Warning> ELSE {4.008} EQ 0 OR 1 AND MO [Integers]		B
Field: 4.008- Compression Algorithm Metadata	8.4.8, Table 24	<The CGA is verified by checking the image metadata for the compression type signature if compression is used.>	2	4.008-Matches Image Metadata	IF {4.008} EQ 1 THEN Present{SOI-WSQ, EOI-WSQ} AND IF {4.008} EQ 2 OR 3 THEN	t-11	B

74

Field	Ref	Char	M	Field name	Condition/Rule	Codes	Type
Field: 4.009-Image Data Valid	8.4.9				This is a mandatory field.		

		2	M	4.009-Uncompressed Image Length	Present(SOI-JPEGB,JPEGL, EOI-JPEG, JPEGL)
					IF {4.008} EQ 0 THEN
					Length{4.009} EQ {4.006} * {4.007}
					ELSE
							B

		2	M	4.009-Valid Image Format	IF {4.008} EQ 1 THEN
					Present(SOI-WSQ,SOF-WSQ,SOB-WSQ,EOI-WSQ)
					ELSE
					IF {4.008} EQ 2 OR 3 THEN
					Present(JFIF,
					SOI-JPEGB,JPEGL,
					SOF-JPEGB,JPEGL,
					EOI-JPEG, JPEGL)
							B

Field: 4.009-Image WSQ Version 2.0	8.4.9, 7.7.9.1	2	M	4.009-Valid WSQ Encoder Version	IF {4.008} EQ 1 THEN {Encoder Version } EQ 1 OR 2	t-11, t-16	B

Wavelet Scalar Quantization (WSQ) shall be used for compressing grayscale friction ridge data at 500 ppi class. Only version 3.1 or higher shall be used for compressing grayscale fingerprint data at 500 ppi class with a platen area of 2 inches or greater in height. WSQ 2.0 or higher may be used for 500 ppi class data taken from a platen of less than 2 inches in height. WSQ shall not be used for other than the 500 ppi class.

Table 6.7 - Assertions for Record Type 5 – Deprecated

Requirement ID	Reference in Base Standard	Level	Status	Assertion ID	Requirement Summary	Test Assertion	Test Note	Implementation Support	Supported Range	Test Result	Applicability
8.5: Record Type-5: DEPRECATED											
Transaction: Type5 Zero Occurrences	8.5	2	M	Type5-Zero Occurrences	No instances of Record Type-5 shall be included in a transaction conformant with this version of the standard.	ForEach(Record in Transaction) { Type(Record) NEQ 5 }					T
		2	M	NIEM-Type5-Zero Occurrences		<An invalid record type will cause a parsing error in XML because no Record Element Tag is defined (see Table 100).>					X

Table 6.8 - Assertions for Record Type 6 – Deprecated

Requirement ID	Reference in Base Standard	Level	Status	Assertion ID	Requirement Summary	Test Assertion	Test Note	Implementation Support	Supported Range	Test Result	Applicability
8.6: Record Type-6: DEPRECATED											
Transaction: Type6 Zero Occurrences	8.6	2	M	Type6-Zero Occurrences	No instances of Record Type-6 shall be included in a transaction conformant with this version of the standard.	ForEach(Record in Transaction) { Type(Record) NEQ 6 }					T
		2	M	NIEM-Type6-Zero Occurrences		<An invalid record type will cause a parsing error in XML because no Record Element Tag is defined (see Table 100).>					X

Table 6.9 - Assertions for Record Type 10 - Facial, Other Body Parts & SMT Image Record

Requirement ID	Reference in Base Standard	Requirement Summary	Lev el	Stat us	Assertion ID	Test Assertion	Test Note	Implementation Support	Supported Range	Test Result	Applicability
		8.10: Record Type-10: Facial, other body part and SMT image record									
Field: Type10-Subfield Occurrence	Table 57	<Table 57 specifies which fields contain subfields as well as the number of occurrences permitted>	1	M	10.[001 to 018, 020, 021, 023, 025, 027, 030, 031, 038, 039, 041, 903, 904, 993, 996, 998, 999]-SubfieldCount	Count(Subfields in 10.[001 to 018, 020, 021,023,025, 027, 030,031,038, 039, 041,903,904,993, 996,998, 999]) EQ 1					T
			1	M	10.[001 to 013, 016, 017, 020, 021, 027, 030, 031, 038, 039, 903, 993, 996, 999]-InfoItemCount	Count(InfoItems in Subfield: 1 in 10.[001 to 013, 016, 017, 020, 021, 027, 030, 031, 038, 039, 903, 993, 996, 999]) EQ 1					T
			1	D	10.014-InfoItemCount	Count(InfoItems in 10.014) EQ 4 OR 5					T
			1	O	10.015-InfoItemCount	Count(InfoItems in 10.015) EQ 2 + 2*(InfoItem:2 in 10.015)					T
			1	D	10.018-InfoItemCount	Count(InfoItems in 10.018) EQ 3					T
			1	D	10.019-SubfieldCount	Count(Subfields in 10.019) MO [1 to 3]					T
			1	D	10.019-	ForEach(Subfield in 10.019)					T

Field	Type	Value	Definition	
InfoItemCount			{ Count(InfoItems in Subfield) EQ 1 }	T
10.023-InfoItemCount	D	1	IF (InfoItem:1 in 10.023) EQ ASCII(VENDOR) THEN Count(InfoItems in 10.023) EQ 1 OR 2 ELSE Count(InfoItems in 10.023) EQ 1	T
10.024-SubfieldCount	D	1	Count(Subfields in 10.024) MO [1 to 9]	T
10.024-InfoItemCount	D	1	ForEach(Subfield in 10.024) { Count(InfoItems in Subfield) EQ 3 }	T
10.025-InfoItemCount	D	1	Count(InfoItems in 10.025) MO [3 to 6]	T
10.026-SubfieldCount	D	1	Count(Subfields in 10.026) MO [1 to 50]	T
10.026-InfoItemCount	D	1	ForEach(Subfield in 10.026) { Count(InfoItems in Subfield) EQ 1 }	T
10.028-SubfieldCount	D	1	Count(Subfields in 10.028) EQ 1 OR 2	T
10.028-InfoItemCount	D	1	ForEach(Subfield in 10.028) { Count(InfoItems in Subfield) EQ 1 }	T
10.029-SubfieldCount	D	1	Count(Subfields in 10.029) MO [1 to 88]	T
10.029-InfoItemCount	D	1	ForEach(Subfield in 10.029) { Count(InfoItems in Subfield) EQ 4 }	T
10.032-SubfieldCount	D	1	Count(Subfields in 10.032) MO [1 to 88]	T
10.032-InfoItemCount	D	1	ForEach(Subfield in 10.032) { Count(InfoItems in Subfield) EQ 5 }	T

			}				
1	D	10.033-SubfieldCount	Count(Subfields in 10.033) MO [1 to 12]				T
1	D	10.033-InfoItemCount	ForEach(Subfield in 10.033) { Count(InfoItems in Subfield) EQ 2 + 2*[InfoItem:2 in Subfield] }				T
1	D	10.040-SubfieldCount	Count(Subfields in 10.040) MO [1 to 3]				T
1	D	10.040-InfoItemCount	ForEach(Subfield in 10.040) { Count(InfoItems in Subfield) EQ 1 }				T
1	D	10.041-InfoItemCount	Count(InfoItems in 10.041) EQ 2				T
1	D	10.042-SubfieldCount	Count(Subfields in 10.042) MO [1 to 9]				T
1	D	10.042-InfoItemCount	ForEach(Subfield in 10.042) { IF [InfoItem:1 in Subfield] MO [ASCII(TATTOO, CHEMICAL, BRANDED, CUT)] THEN Count(InfoItems in Subfield) MO [3,4] ELSE Count(InfoItems in Subfield) EQ 1 }				T
2	D	10.043-SubfieldCount	Count(Subfields in 10.043) MO [1 to 9]				T
1	D	10.043-InfoItemCount	ForEach(Subfield in 10.043) { Count(InfoItems in Subfield) MO [1 to 6] }				T
1	O	10.044-SubfieldCount	Count(Subfields in 10.044) MO [1 to 18]				T
1	O	10.044-InfoItemCount	ForEach(Subfield in 10.019) { Count(InfoItems in Subfield) EQ 1 }				T

Type	Count		Field	Condition	
T	2	D	10.045-SubfieldCount	Count(Subfields in 10.045) MO [1 to 16]	
T	1	D	10.045-InfoItemCount	ForEach(Subfield in 10.045) { Count(InfoItems inSubfield) EQ 3 + 2*{ InfoItem:3 in Subfield} }	
T	1	O	10.902-SubfieldCount	Count(Subfields in 10.902) GTE 1	
T	1	O	10.902-InfoItemCount	ForEach(Subfield in 10.902) { Count(InfoItems in Subfield) EQ 4 }	
T	1	O	10.904-InfoItemCount	Count(InfoItems in 10.904) EQ 3	
T	1	O	10.995-SubfieldCount	Count(Subfields in 10.995) MO [1 to 255]	
T	1	O	10.995-InfoItemCount	ForEach(Subfield in 10.995) { Count(InfoItems in Subfield) EQ 1 OR 2 }	
T	1	O	10.997-SubfieldCount	Count(Subfields in 10.997) MO [1 to 255]	
T	1	O	10.997-InfoItemCount	ForEach(Subfield in 10.997) { Count(InfoItems in Subfield) EQ 1 OR 2 }	
	1	O	10.998-SubfieldCount	<See Requirement ID: "Field: Geographic">	t-2
B	1	M	10.[001 to 012, 999]-Mandatory CondCode	Present(10.001 to 10.012, 10.999)	
B	1	M	[10.034 to 10.037,10. 046 to 10.199,	NOT Present(10.034 to 10.037,10.046 to 10.199, 10.901,10.905 to 10.992, 10.994)	

Field: Type10-CondCode

Table 57 <Table 57 specifies the Condition Code for each field.>

80

Field	Table Ref		Char	Code	CondCode Field	Condition	Reserved						Type
							10.901, 10.905 to 10.994]-Reserved						
						<See Requirement ID: "Field: SAP Conditional">	t-2						
Field: 10.013-SAP CondCode Dependent	Table 57, 8.10.13		D		10.013-CondCode Dependent	The Subject Acquisition Profile (SAP) is a mandatory field when Field 10.003: Image type / IMT contains "FACE". Otherwise, it shall not be entered.							B
Field: 10.014-FIP CondCode Dependent	Table 57, 8.10.14	2	D		10.014-CondCode Dependent	IF Present(10.014) THEN {10.003} EQ ASCII(FACE) AND {10.013} NOT MO [30, 32, 40, 42, 50, 51, 52]	This field is only appropriate for face images (IMT = 'FACE') that do not comply with SAP Levels 30, 32, 40, 42, 50, 51 or 52, because those images shall be cropped to a "head only" or "head and shoulders" composition.						B
Field: 10.015-FPFi CondCode Dependent	Table 57, 8.10.15	2	D		10.015-CondCode Dependent	IF Present(10.015) THEN {10.003} EQ ASCII(FACE) AND {10.013} NOT MO [30, 32, 40, 42, 50, 51, 52]	This field is only appropriate for images that do not comply with SAP Levels 30, 32, 40, 42, 50, 51 or 52, because those images shall be cropped to a "head only" or "head and shoulders" composition.						B
Field: 10.018-DIST CondCode Dependent	Table 57, 8.10.18	2	D		10.018-CondCode Dependent	IF Present(10.018) THEN {10.003} EQ ASCII(FACE)	This optional field (which can be used only if IMT is 'FACE')...						B
Field: 10.019-LAF CondCode Dependent	Table 57, 8.10.19	2	D		10.019-CondCode Dependent	IF Present(10.019) THEN {10.003} EQ ASCII(FACE)	This optional field ...is only applicable to face images (IMT = 'FACE').						B
Field: 10.020-POS CondCode Dependent	Table 57, 8.10.20	2	D		10.020-CondCode Dependent	IF Present(10.020) THEN {10.003} EQ ASCII(FACE)	This optional field is to be used for the exchange of facial image data						B
Field: 10.021-POA CondCode Dependent	Table 57, 8.10.21	2	D		10.021-CondCode Dependent	IF Present(10.021) THEN {10.003} EQ ASCII(FACE) AND {10.020} EQ ASCII(A)	This shall only be used for the exchange of facial image data (IMT = 'FACE'). It may be used if Field 10.020: Subject pose / POS contains an "A" to indicate an angled pose of the subject. The field shall not be used if the entry in POS is an "F", "R", "L" or "D".						
Field: 10.023-PAS CondCode Dependent	Table 57, 8.10.22	2	D		10.023-CondCode Dependent	IF {10.013} GTE 40 THEN Present(10.023)	This field is mandatory if the SAP entry (Field 10.013: Subject acquisition profile / SAP) is "40" or greater for face image records. (IMT=FACE only).						B
Field: 10.024-SQS CondCode Dependent	Table 57, 8.10.23	2	D		10.024-CondCode Dependent	IF Present(10.024) THEN {10.003} EQ ASCII(FACE)	This optional field shall specify quality score data for facial images (IMT = 'FACE')						B

Field	Reference	Description			CondCode	Condition	
Field: 10.025-SPA CondCode Dependent	Table 57, 8.10.24	This field shall be present when Field 10.020: Subject pose / POS contains a "D" to indicate a set of determined 3D pose angles of the same subject for a facial image (IMT = 'FACE'). If the entry in POS is an "F", "L", or "R" this field shall not be used.	2	D	10.025-CondCode Dependent	IF {10.020} EQ ASCII(D) THEN Present(10.025) AND IF {10.020} MO[F,L,R] THEN Not Present(10.025) AND IF Present(10.025) THEN {10.003} EQ ASCII(FACE)	B
Field: 10.026-SXS CondCode Dependent	Table 57, 8.10.25	This field is mandatory if the SAP entry for a facial image (Field 10.013: Subject acquisition profile / SAP) is 40, 50 or 51. (IMT=FACE only). In other cases, this field is optional for facial images.	2	D	10.026-CondCode Dependent	IF {10.013} MO [40,50,51] AND {10.003} EQ ASCII(FACE) THEN Present(10.026) AND IF Present(10.026) THEN {10.003} EQ ASCII(FACE)	B
Field: 10.027-SEC CondCode Dependent	Table 57, 8.10.26	This field is mandatory if the SAP entry (Field 10.013: Subject acquisition profile / SAP) is "40" or greater. For other facial images (IMT=FACE), the field is optional.	2	D	10.027-CondCode Dependent	IF {10.013} GTE 40 AND {10.003} EQ ASCII(FACE) THEN Present(10.027) AND IF Present(10.026) THEN {10.003} EQ ASCII(FACE)	B
Field: 10.028-SHC CondCode Dependent	Table 57, 8.10.27	This field is mandatory if the SAP entry (Field 10.013: Subject acquisition profile / SAP) is "40" or greater. For other facial images (IMT ='FACE'), it is optional.	2	D	10.028-CondCode Dependent	IF {10.013} GTE 40 AND {10.003} EQ ASCII(FACE) THEN Present(10.028) AND IF Present(10.026) THEN {10.003} EQ ASCII(FACE)	B
Field: 10.029-FFP CondCode Dependent	Table 57, 8.10.28	The optional field shall be used for the exchange of facial image data (IMT = 'FACE') feature points or landmarks.	2	D	10.029-CondCode Dependent	IF Present(10.029) THEN {10.003} EQ ASCII(FACE)	B
Field: 10.031-TMC CondCode Dependent	Table 57, 8.10.30	This optional field describes the specific facial (IMT = 'FACE') feature points	2	D	10.031-CondCode Dependent	IF Present(10.031) THEN {10.003} EQ ASCII(FACE)	B
Field: 10.032-3DF CondCode Dependent	Table 57, 8.10.31	The optional field shall describe ...facial feature points of the captured facial image(IMT ='FACE').	2	D	10.032-CondCode Dependent	IF Present(10.032) THEN {10.003} EQ ASCII(FACE)	B
Field: 10.033-FEC	Table 57, 8.10.32,	Field 10.031: This optional field describes the specific facial (IMT= 'FACE') feature	2	D	10.033-CondCode	Present(10.033) IFF {10.003} EQ ASCII(FACE) AND {10.031} EQ 5	B

Field	Reference	Description	Level	M/D	Element / CharType	Condition	
CondCode Dependent	8.10.30	points contained in Field 10.029: 2d Facial feature points/ FFP and if level 5, contours shall be contained in Field 10.033: Feature contours/ FEC.			Dependent		
Field: 10.039-T10 CondCode Dependent	Table 57, 8.10.34	This field shall only be present if multiple Type-10 records in the transaction contain the same SMT or body part.	2	D	10.039-CondCode Dependent	IF Present(10.039) THEN Count(Records ST Type(Record) EQ 10) GTE 2	T
			2	M	NIEM- T10-CondCode Dependent	IF Present(XEIm(biom:PhysicalFeatureReferenceIde ntification) THEN Count(XEIm(itl:PackageFacialAndSMTImageRecor d)) GTE 2	X
Field: 10.040-SMT CondCode Dependent	Table 57, 8.10.35	This field shall be used only when Field 10.003: Image type / IMT = "SCAR", "MARK", or "TATTOO". It is not used for other images	2	D	10.040-CondCode Dependent	IF Present(10.040) THEN {10.003} MO [ASCII(SCAR,MARK, TATTOO)]	B
Field: 10.041-SMS CondCode Dependent	Table 57, 8.10.36	This field shall be used only when Field 10.003: Image type / IMT = "SCAR", "MARK", or "TATTOO".	2	D	10.041-CondCode Dependent	IF Present(10.041) THEN {10.003} NEQ ASCII(FACE)	B
Field: 10.042-SMD CondCode Dependent	Table 57, 8.10.37	This field shall be used only when Field 10.003: Image type / IMT = "SCAR", "MARK", or "TATTOO".	2	D	10.042-CondCode Dependent	IF Present(10.042) THEN {10.003} MO [ASCII(SCAR,MARK, TATTOO)]	B
Field: 10.043-COL CondCode Dependent	Table 57, 8.10.38	It shall contain one subfield corresponding to each subfield contained in Field 10.042: SMT descriptors / SMD	2	D	10.043-CondCode Dependent	IF Present(10.043) THEN Present(10.042)	B
Field: 10.045-OCC CondCode Dependent	Table 57, 8.10.40	This optional field defines ...the image of the face (IMT = 'FACE').	2	D	10.045-CondCode Dependent	IF Present(10.045) THEN {10.003} EQ ASCII(FACE)	B
Field: Type10-CharType	8.10, Table 57	<Table 57 specifies the Character Type for each field that contains no subfields..>	1	-	10.[001,00 2,006 to 010, 013,016,01 7,031,039]-CharType	Bytes(10.[001,002,006 to 010, 013,016,017,031,039]) MO [0x30 to 0x39]	B
			1	-	10.[012,02 0,027,030] -CharType	Bytes(10.[012,020,027,030]) MO [0x20, 0x41 to 0x5A, 0x61 to 0x7A]	
			1	M	10.003-CharType	Bytes(10.003) MO [0x2D, 0x20, 0x41 to 0x5A, 0x61 to 0x7A]	B
			1	M	10.004-CharType	<See Requirement ID: "Field: Originating Agency".>	B

t-2

Field	Ref	Occ	Cond	Subfield	Description		Char Type
		1	M	10.005-CharType	Bytes(10.005) MO [0x30 to 0x39]		T
		1	M	NIEM-10.005-CharType	Bytes(10.005) MO [0x30 to 0x39, 0x2D]		X
		1	M	10.011-CharType	Bytes(10.011) MO [0x20, 0x41 to 0x5A, 0x61 to 0x7A, 0x30 to 0x39]		T
		1	D	10.021-CharType	Bytes(10.021) MO [0x2D,0x30 to 0x39]		B
		1	O	10.038-CharType	TRUE		B
		1	O	10.903-CharType	Bytes(10 903) MO [0x20 to 0x7E]		B
		1	M	10.993-CharType	<See Requirement ID: "Field: Source Agency Name".>	t-2	
		1	O	10.996-CharType	Bytes(10 996) MO [0x30 to 0x39,0x41 to 0x46, 0x61 to 0x66]		B
		1	M	10.999-CharType	TRUE		B
Field: Type10-Subfield CharType	8.10, Table 57	1	D	10.014-[LHC, RHC, TVC, BVC]-CharType	Bytes(InfoItem:1 to 4 in 10.014) MO [0x30 to 0x39]		B*
		1	O	10.014-BBC-CharType	Bytes(InfoItem:5 in 10.014) MO [0x20, 0x41 to 0x5A, 0x61 to 0x7A]		B*
		1	O	10.015-BYC-CharType	Bytes(InfoItem:1 in 10.015) MO [0x20, 0x41 to 0x5A, 0x61 to 0x7A]		B*
		1	O	10.015-[NOP, HPO, VPO]-CharType	Bytes(All(InfoItems in 10.015 ST InfoItems NOT InfoItem:1 in 10.015)) MO [0x30 to 0x39]		B*
		1	M	10.018-[IDK, IDM, DSC]-CharType	Bytes(All(InfoItems in 10.018)) MO [0x20, 0x41 to 0x5A, 0x61 to 0x7A]		B*
		1	D	10.019-LAF-CharType	Bytes(All(InfoItems in 10.019)) MO [0x20, 0x41 to 0x5A, 0x61 to 0x7A]		B*
		1	D	10.023-PAC-CharType	Bytes(InfoItem:1 in 10.023) MO [0x20, 0x41 to 0x7A, 0x30 to 0x39]		B*
		1	D	10.023-VSD-CharType	TRUE		B*

<Table 57 specifies the Character Type for each subfield.>

				B*
1	D	10.024-[QVU, QAP]-CharType	ForEach(Subfield in 10.024) { Bytes(InfoItem:1,3 in Subfield) MO [0x30 to 0x39] }	B*
1	D	10.024-QAV-CharType	ForEach(Subfield in 10.024) { Bytes(InfoItem:2 in Subfield) MO [0x30 to 0x39,0x41 to 0x46, 0x61 to 0x66] }	B*
1	D	10.025-[YAW, PIT, ROL]-CharType	Bytes(InfoItems: 1 to 3 in 10.025) MO [0x2D, 0x30 to 0x39]	B*
1	D	10.025-[YAWU, PITU, ROLU]-CharType	Bytes(InfoItems: 4 to 6 in 10.025) MO [0x30 to 0x39]	B*
1	D	10.026-SXS-CharType	Bytes(All(InfoItems in 10.018)) MO [0x20, 0x41 to 0x5A, 0x61 to 0x7A]	B*
1	D	10.028-SHC-CharType	Bytes(All(InfoItems in 10.028)) MO [0x20, 0x41 to 0x5A, 0x61 to 0x7A]	B*
1	D	10.029-[FPT, HCX, HCY]-CharType	ForEach(Subfield in 10.029) { Bytes(InfoItem:1,3,4 in Subfield) MO [0x30 to 0x39] }	B*
1	D	10.029-FPC-CharType	ForEach(Subfield in 10.029) { Bytes(InfoItem:2 in Subfield) MO [0x2E, 0x30 to 0x39,0x61 to 0x7A] }	B*
1	D	10.032-[FPT, HCX, HCY, HCZ]-CharType	ForEach(Subfield in 10.032) { Bytes(InfoItem:1,3,4,5 in Subfield) MO [0x30 to 0x39] }	B*
1	D	10.032-FPC-CharType	ForEach(Subfield in 10.032) { Bytes(InfoItem:2 in Subfield) MO [0x2E, 0x30 to 0x39,0x61 to 0x7A] }	B*
1	D	10.033-	ForEach(Subfield in 10.033)	B*

		Field	Expression	Type
		FCC CharType	Bytes(InfoItem:1 in Subfield) MO [0x20, 0x41 to 0x5A, 0x61 to 0x7A]	B*
1	D	10.033-[NOP, HPO, VPO] CharType	ForEach(Subfield in 10.033) { Bytes(All(InfoItems in Subfield ST InfoItems NOT InfoItem:1 in Subfield)) MO [0x30 to 0x39] }	B*
1	D	10.040-SMT-CharType	Bytes(All(InfoItems in 10.040)) MO [0x20, 0x41 to 0x5A, 0x61 to 0x7A]	B*
1	D	10.041-[HGT, WID]-CharType	Bytes(All(InfoItems in 10.041)) MO [0x30 to 0x39]	B*
1	D	10.042-[SMI, TAC, TSC]-CharType	ForEach(Subfield in 10.042) { Bytes(InfoItem:1 to 3 in Subfield) MO [0x20, 0x41 to 0x46, 0x61 to 0x66] }	B*
1	D	10.042-TDS-CharType	TRUE	B*
1	D	10.043-[TC1, TC2, TC3, TC4, TC5, TC6]-CharType	Bytes(All(InfoItems in 10.043)) MO [0x20, 0x41 to 0x46, 0x61 to 0x66]	B*
1	O	10.044-ITX-CharType	Bytes(All(InfoItems in 10.044)) MO [0x20, 0x41 to 0x46, 0x61 to 0x66]	B*
1	D	10.045-[OCY, OCT]-CharType	ForEach(Subfield in 10.045) { Bytes(InfoItem:1,2 in Subfield) MO [0x20, 0x41 to 0x5A, 0x61 to 0x7A] }	B*
1	D	10.045-[NOP, HPO, VPO]-CharType	ForEach(Subfield in 10.045) { Bytes(All(InfoItems in Subfield ST InfoItems NOT InfoItem:1 OR InfoItem:2 in Subfield)) MO [0x30 to 0x39] }	B*
1	O	10.902-[NAV, OWN,	TRUE	T

Field	Ref	Count	M/O	Element	Rule	Req	Type
				PRO]-CharType	ForEach(Subfield in 10.902)		T
		1	O	10.902-[GMT]-CharType	Bytes(Infoitem:1 in Subfield) MO [0x30 to 0x39,0x5A] }		X*
		1	O	NIEM-10.902-Subfield CharType	< The treatment of subfields for validation in the XML version requires further review. Byte values allowed for first "subfield" in XML are 0x30 to 0x39, 0x3A, 0x54, 0x5A.>		B*
		1	O	10.904-[MAK, MOD, SER]-CharType	TRUE		B*
		1	O	10.995-[ACN, ASP]-CharType	Bytes(All(Infoitems in 10.995)) MO [0x30 to 0x39]		B*
		1	O	10.997-[SRN, RSP]-CharType	Bytes(All(Infoitems in 10.997)) MO [0x30 to 0x39]		B*
		1	O	10.998-[UTE, LTD, LTM, LTS, LGD, LGM, LGS, ELE, GDC, GCM, GCE, GCN, GRT, OSI, OCV]-CharType	<See Requirement ID: "Field: Geographic">	t-2	
Field: Type10-CharCount	Table 57, 7.1	1	M	10.001-CharCount	DataLength(10.001) MO [1 to 8]		T
	<Table 57 specifies the Character Count for each field that contains no subfields.>	1	M	NIEM-10.001-CharCount	Length(10.001) EQ 2		X
		1	M	10.002-CharCount	DataLength(10.002) EQ 1 OR 2		B
		1	M	10.003-CharCount	DataLength(10.003) MO [4 to 11]		B
		1	M	10.004-CharCount	<See Requirement ID: "Field: Originating Agency".>	t-2	T
		1	M	10.005-CharCount	DataLength(10.005) EQ 8		
		1	M	NIEM-10.005-	DataLength(10.005) EQ 10		X

1	M	CharCount		
1	M	10.006-CharCount	DataLength(10.006) MO [2 to 5]	B
1	M	10.007-CharCount	DataLength(10.007) MO [2 to 5]	B
1	M	10.008-CharCount	DataLength(10.008) EQ 1	B
1	M	10.009-CharCount	DataLength(10.009) MO [1 to 5]	B
1	M	10.010-CharCount	DataLength(10.010) MO [1 to 5]	B
1	M	10.011-CharCount	DataLength(10.011) MO [3 to 5]	B
1	M	10.012-CharCount	DataLength(10.012) MO [3 to 4]	B
1	D	10.013-CharCount	DataLength(10.013) MO [1 to 2]	B
1	O	10.016-CharCount	DataLength(10.016) MO [1 to 5]	B
1	O	10.017-CharCount	DataLength(10.017) MO [1 to 5]	B
1	D	10.019-CharCount	DataLength(10.019) MO [1 to 3]	B
1	D	10.020-CharCount	DataLength(10.020) EQ 1	B
1	D	10.021-CharCount	DataLength(10.021) MO [1 to 4]	B
1	D	10.027-CharCount	DataLength(10.027) EQ 3	B
1	O	10.030-CharCount	DataLength(10.030) MO [7 to 10]	B
1	D	10.031-CharCount	DataLength(10.031) MO [1 to 3]	B
1	O	10.038-CharCount	DataLength(10.038) MO [1 to 126]	B
1	D	10.039-CharCount	DataLength(10.039) MO [1 to 3]	B
1	O	10.903-CharCount	DataLength(10.903) MO [13 to 16]	B
1	O	10.993-CharCount	<See Requirment ID: "Field: Source Agency Name".>	t-2
1	O	10.996-CharCount	DataLength(10.996) EQ 64	B

Field:				
Field: Type 10- Subfield CharCount		O	10.998- CharCount	<See Requirement ID: "Field: Geographic"> t-2	B*

Table 57	<Table 57 specifies the Character Count for each subfield.>			
		1	D	10.014- [LHC, RHC, TVC, BVC]- CharCount	Length(InfoItem:1 to 4 in 10.014) MO [1 to 5]	B*
		1	D	10.014- [BBC]- CharCount	Length(InfoItem:5 in 10.014) EQ 1	B*
		1	O	10.015- BYC- CharCount	Length(InfoItem:1 in 10.015) EQ 1	B*
		1	O	10.015- NOP- CharCount	Length(InfoItem:2 in 10.015) EQ 1 OR 2	B*
		1	O	10.015-[HPO, VPO]- CharCount	Length(All((InfoItems in 10.015 ST InfoItems NOT InfoItem:1 OR InfoItem 2 in 10.015) MO [1 to 5]	B*
		1	D	10.018- IDK- CharCount	Length(InfoItem:1 in 10.018) MO [6 to 10]	B*
		1	D	10.018- IDM- CharCount	Length(InfoItem:2 in 10.018) EQ 1	B*
		1	D	10.018- DSC- CharCount	Length(Last(InfoItem in 10.018)) MO [4 to 8]	B*
		1	D	10.019- LAF- CharCount	Length(All((InfoItems in 10.019) EQ 1	B*
		1	D	10.023- PAC- CharCount	Length(InfoItem:1 in 10.023) MO [6 to 14]	B*
		1	D	10.023- VSD- CharCount	Length(InfoItem:2 in 10.023) MO [1 to 64]	B*
		1	D	10.024- QVU- CharCount	ForEach(Subfield in 10.024) { Length(InfoItem:1 in Subfield) MO [1 to 3] }	B*
		1	D	10.024- QAV- CharCount	ForEach(Subfield in 10.024) { Length(InfoItem:2 in Subfield) EQ 4 }	B*
		1	D	10.024- QAP-	ForEach(Subfield in 10.024) {	B*

		CharCount	Length(InfoItem:3 in Subfield)) MO [1 to 5] }	B*
1	D	10.026-SXS-CharCount	Length(All(InfoItems in 10.026)) MO [3 to 20]	B*
1	D	10.028-SHC-CharCount	Length(All(InfoItems in 10.028)) EQ 3	B*
1	D	10.029-FPT-CharCount	ForEach(Subfield in 10.029) { Length(InfoItem:1 in Subfield) EQ 1	B*
1	D	10.029-FPC-CharCount	ForEach(Subfield in 10.029) { Length(InfoItem:2 in Subfield) MO [3 to 5] }	B*
1	D	10.029-[HCX, HCY]-CharCount	ForEach(Subfield in 10.029) { Length(InfoItem:3,4 in Subfield)) MO [1 to 5]	B*
1	D	10.032-FPT-CharCount	ForEach(Subfield in 10.032) { Length(InfoItem:1 in Subfield) EQ 1	B*
1	D	10.032-FPC-CharCount	ForEach(Subfield in 10.032) { Length(InfoItem:2 in Subfield) MO [3, 5] }	B*
1	D	10.032-[HCX, HCY, HCZ]-CharCount	ForEach(Subfield in 10.032) { Length(InfoItem:3 to 5 in Subfield)) MO [1 to 5]	B*
1	D	10.033-FCC-CharCount	ForEach(Subfield in 10.033) { Length(InfoItem:1 in Subfield) MO [4 to 14]	B*
1	D	10.033-NOP-CharCount	ForEach(Subfield in 10.033) { Length(InfoItem:2 in Subfield) EQ 1 OR 2	B*
1	D	10.033-[HPO, VPO]-CharCount	ForEach(Subfield in 10.033) { Length((All(InfoItems in Subfield ST InfoItems NOT InfoItem::1 OR InfoItem:2 in Subfield) MO [1 to 5] }	B*

1	D	10.040-SMT-CharCount	Length(All(InfoItems in 10.040)) MO [3 to 10]	B*
1	D	10.041-[HGT, WID]-CharCount	Length(InfoItem:1,2 in 10.041)) MO [1 to 3]	B*
1	D	10.042-SMI-CharCount	ForEach(Subfield in 10.042) { Length(InfoItem:1 in Subfield) MO [3 to 8] }	B*
1	D	10.042-TAC-CharCount	ForEach(Subfield in 10.042) { Length(InfoItem:2 in Subfield) MO [4 to 8] }	B*
1	D	10.042-TSC-CharCount	ForEach(Subfield in 10.042) { Length(InfoItem:3 in Subfield)) MO [3 to 9] }	B*
1	D	10.042-TDS-CharCount	ForEach(Subfield in 10.042) { Length(InfoItem:4 in Subfield)) MO [1 to 256] }	B*
1	D	10.043-[TC1, TC2, TC3, TC4, TC5, TC6]-CharCount	Length(All(InfoItems in 10.043)) MO [3 to 7]	B*
1	O	10.044-ITX-CharCount	Length(All(InfoItems in 10.044)) MO [3 to 11]	B*
1	D	10.045-[OCY, OCT]-CharCount	ForEach(Subfield in 10.045) { Length(InfoItem:1,2 in Subfield) EQ 1 }	B*
1	D	10.045-NOP-CharCount	ForEach(Subfield in 10.045) { Length(InfoItem:3 in Subfield) EQ 1 OR 2 }	B*
1	D	10.045-[HPO, VPO]-CharCount	ForEach(Subfield in 10.045) { Length((All(InfoItems in Subfield ST InfoItems NOT InfoItem:1 OR InfoItem:2 OR InfoItem:3 in Subfield) MO [1 to 5] }	B*
1	O	10.902-	ForEach(Subfield in 10.902)	T

Occurrence	M/O	Code	Condition		Type
		GMT-CharCount	{ Length(InfoItem:1 in Subfield) EQ 15 }		T
1	O	10.902-[NAV, OWN]-CharCount	ForEach(Subfield in 10.902) { Length(InfoItem:2,3 in Subfield) MO [1 to 64] }		T
1	O	10.902-PRO-CharCount	ForEach(Subfield in 10.902) { Length(InfoItem:4 in Subfield)) MO [1 to 255] }		
1	O	NIEM-10.902-Subfield CharCount	< The treatment of subfields for validation in the XML version requires further review. Length of the first "subfield" in XML is 20.>		X*
1	O	10.904-[MAK, MOD, SER]-CharCount	Length(All(InfoItems in 10.904)) MO [1 to 50]		B*
1	O	10.995-ACN-CharCount	ForEach(Subfield in 10.995) { Length(InfoItem:1 in Subfield) MO [1 to 3] }		B*
1	O	10.995-ASP-CharCount	ForEach(Subfield in 10.995) { Length(InfoItem:2 in Subfield) EQ 1 OR 2 }		B*
1	O	10.997-SRN-CharCount	ForEach(Subfield in 10.997) { Length(InfoItem:1 in Subfield) MO [1 to 3] }		B*
1	O	10.997-RSP-CharCount	ForEach(Subfield in 10.997) { Length(InfoItem:2 in Subfield) EQ 1 OR 2 }		B*
1	O	10.998-Subfield CharCount	<See Requirement ID: "Field: Geographic" >	t-2	
1 2	-	10.[022, 046 to 199,901,905 to 994]-Occurrence	Count(10.[034 to 037, 046 to 199,901,905 to 992, 99]) EQ 0		B
1	M	10.[001 to 012, 999]-Occurrence	Count(10.[001 to 012, 999]) EQ 1		B

Table 57 — <Table 57 specifies the Field Occurrence for each field.>

Field: Type10-Field Occurrence

92

Field	Reference	Description	Min	M	Value ID	Content	Type	Code
			1	-	10.[013 to 021,023 to 033,038 to 045, 902 to 904, 993, 995 to 998]-Occurrence	Count(10.[013 to 021,023 to 033,038 to 045, 902 to 904, 993, 995 to 998]) LTE 1		B
			1	M	10.022-Occurrence Legacy	IF Present (10.022) <Provide Legacy Warning>.		B
Field: 10.001-Record Header Value	8.10.1, Table 57, 7.1	Field 10.001 Record header. In Traditional encoding, this field contains the record length in bytes (including all information separators)	1	M	10.001-Record Header	<See Requirement ID "Field: xx.001-Record Header">	t-2	
	8.10.1, C.9.8	The XML name for the Type-10 record is <itl:PackageFacialAndSMTImageRecord>, and its <biom:RecordCategoryCode> element shall have a value of "10".	1	M	NIEM-10.001-Value	ForEach(itl:PackageFacialAndSMTImageRecord) {XElm(biom:RecordCategoryCode) EQ ASCII(10)}		X
Field: 10.002-Information Designation Character Value	8.10.2, Table 57, 7.3.1	This mandatory field shall contain the IDC assigned to this Type-10 record as listed in the information item IDC for this record in Field 1.003 Transaction content/CNT.	1	M	10.002-Value	<See Requirement IDs "Field: xx.002-IDC" and "Field: 1.003-Transaction Content Subfield 2 IDC Matches">	t-2	
Field: 10.003-Image Type Value	8.10.3, Table 57, Table 58	This mandatory field shall be used to indicate the type of image contained in this record. It shall contain a character string from the "Image Code" column of Table 58 to indicate the appropriate image type.	1	M	10.003-Value	{10.003} MO [ASCII(SCAR, TATTOO, FACE, FRONTAL-C, REAR-C, FRONTAL-N, REAR-N, TORSO-BACK, TORSO-FRONT, CONDITION, MISSING, OTHER, CHEST, FEET, HANDS-PALM, HANDS-BACK, GENITALS, BUTTOCKS, RIGHT LEG, LEFT LEG, RIGHT ARM, LEFT ARM, MARK)]	t-1	B
Field: 10.004-Source Agency Value	8.10.4, 7.6	The data content of this field is defined by the user and shall be in accordance with the receiving agency.	1	M	10.004-Value	<See Requirement ID: "Field: Originating Agency".>	t-2	
Field: 10.005-Photo Capture Date Value	8.10.5, 7.7.2.3	This mandatory field shall contain the date that the image contained in the record was captured.	1	M	10.005-Value	{10.005} MO [ValidLocalDate]	t-6	T
			1	M	NIEM-10.005-Value	ForEach(XElm(itl:PackageFacialAndSMTImageRecord)) {XElm(nc:Date) in XElm(biom:CaptureDate)} MO [NIEM-ValidLocalDate]	t-6	X

Field	Reference		M/O	Description	Requirement ID	t-code	
Field: 10.006-Horizontal Line Length Value	8.10.6, Table 57, 7.7.8.1		M	The maximum horizontal size is limited to 65,534 pixels in Record Types-4 and 8, and to 99,999 for other record types. The minimum value is 10 pixels.	<See Requirement ID "Field: Image HLL Value" >	t-2	
Field: 10.006-Horizontal Line Length Metadata	8.10.6, Table 57, 7.7.8.1	2	M	<The HLL is verified by checking the image metadata if compression is used.>	IF {10.011} EQ ASCII(JPEGB) OR ASCII(JPEGL) THEN {10.006} EQ {ImageWidth-JPEGB,JPEGL} ELSE IF {10.011} EQ ASCII(JP2) OR ASCII(JP2L) THEN {10.006} EQ {ImageWidth-JP2,JP2L} ELSE IF {10.011} EQ ASCII(PNG) THEN {10.006} EQ {ImageWidth-PNG}	t-11	B
Field: 10.007-Vertical Line Length Value	8.10.7, Table 57, 7.7.8.2		M	The maximum vertical size is limited to 65,534 pixels in Record Types-4 and 8, and to 99,999 for other record types. The minimum value is 10 pixels.	<See Requirement ID "Field: Image VLL Value" >	t-2	
Field: 10.007-Vertical Line Length Metadata	8.10.7, Table 57, 7.7.8.2	2	M	<The VLL is verified by checking the image metadata if compression is used.>	IF {10.011} EQ ASCII(JPEGB) OR ASCII(JPEGL) THEN {10.007} EQ {ImageHeight-JPEGB,JPEGL} ELSE IF {10.011} EQ ASCII(JP2) OR ASCII(JP2L) THEN {10.007} EQ {ImageHeight-JP2,JP2L} ELSE IF {10.011} EQ ASCII(PNG) THEN {10.007} EQ {ImageHeight-PNG}	t-11	B
Field: 10.008-Scale Units Value	8.10.8, Table 57, 7.7.8.3		M	<Table 57 lists the value constraints for SLC>	<See Requirement ID "Field: Image SLC Value" >	t-2	
Field: 10.008-Scale Units Metadata	8.10.8, Table 57, 7.7.8.3	2	M	A value of "1" shall indicate pixels per inch. A value of "2" shall indicate pixels per centimeter. A value of "0" in this field indicates that no scale is provided, and the quotient of THPS/TVPS shall provide the pixel aspect ratio. <The SLC is verified by checking the image metadata if compression is used.>	IF {10.011} EQ ASCII(JPEGB) OR ASCII(JPEGL) THEN {10.008} EQ {SamplingUnits-JPEGB,JPEGL} ELSE IF {10.011} EQ ASCII(JP2) OR ASCII(JP2) THEN <Provide Warning "Not Tested"> ELSE IF {10.011} EQ ASCII(PNG) THEN IF {10.008} EQ 1 OR 2 THEN {SamplingUnits-PNG} EQ 1, ELSE IF {10.008} EQ 0 THEN {SamplingUnits-PNG} EQ 0	t-11	B
Field: 10.009-Transmitted Horizontal Pixel Scale	8.10.9, Table 57, 7.7.8.4		M	<Table 57 lists the value constraints for THPS.>	<See Requirement ID "Field: Image THPS Value" >	t-2	

Field	Reference		M	Description	Requirement	Value / Logic	Test	
Value								B
Field: 10.009-Transmitted Horizontal Pixel Scale Metadata	8.10.9, Table 57, 7.7.8.4	2	M	This is the integer pixel density used in the horizontal direction of the image if SLC has a value of "1" or "2". If SLC has a value of "0", this information item shall contain the horizontal component of the pixel aspect ratio, up to 5 digits. <The THPS is verified by checking the image metadata if compression is usec >	10.009-Matches Image Metatdata	IF {10.011} EQ ASCII(JPEGB) OR ASCII(JPEGL) AND {10.008} EQ 1 OR 2 THEN {10.009} EQ {HorizontalDensity-JPEGB,JPEGL} ELSE IF {10.011} EQ ASCII(JP2) OR ASCII(JP2L) AND {10.008} EQ 1 OR 2 THEN <Provide Warning "Not Tested"> ELSE IF {10.011} EQ ASCII(PNG) AND {10.008} EQ 1 THEN {10.009} EQ {HorizontalDensity-PNG} * 0.0254 (meters/inch), ELSE IF 10.011} EQ ASCII(PNG) AND {10.008} EQ 2 THEN {10.009} EQ {HorizontalDensity-PNG} * 0.01 (meters/cm)	t-11, t-12	B
		2	M		10.009-Aspect Ratio Matches Metadata	IF {10.011} EQ ASCII(JPEGB) OR ASCII(JPEGL) AND {10.008} NEQ 1 OR 2 THEN {10.009}/{10.010} EQ {HorizontalDensity-JPEGB,JPEGL} / {VerticalDensity-JPEGB,JPEGL} ELSE IF {10.011} EQ ASCII(JP2) OR ASCII(JP2L) AND {10.008} NEQ 1 OR 2 THEN <Provide Warning "Not Tested"> ELSE IF {10.011} EQ ASCII(PNG) AND {10.008} NEQ 1 OR 2 THEN {10.009}/{10.010} EQ {HorizontalDensity-PNG}/ {VerticalDensity-PNG}	t-11	
Field: 10.010-Transmitted Vertical Pixel Scale Value	8.10.10, Table 57, 7.7.8.5	2	M	Table 57 lists the value constraints for TVPS.>	10.010-Value	<See Requirement ID "Field: Image TVPS Value">	t-2	
Field: 10.010-Transmitted Vertical Pixel Scale Metadata	8.10.10, Table 57, 7.7.8.5	2	M	This is the integer pixel density used in the Vertical direction of the image if SLC has a value of "1" or "2". If SLC has a value of "0", this information item shall contain the Vertical component of the pixel aspect ratio, up to 5 digits. <The TVPS is verified by checking the image metadata if compression is used>	10.010-Matches Metadata	IF {10.011} EQ ASCII(JPEGB) OR ASCII(JPEGL) AND {10.008} EQ 1 OR 2 THEN {10.010} EQ {VerticalDensity-JPEGB,JPEGL} ELSE IF {10.011} EQ ASCII(JP2) OR ASCII(JP2L) THEN <Provide Warning "Not Tested"> ELSE IF 10.011} EQ ASCII(PNG) AND {10.008} EQ 1 THEN {10.010} EQ {VerticalDensity-PNG} * 0.0254 (meters/inch),	t-11, t-12	B

95

Field	Reference	Description		M	Requirement ID	Logic	t	B
			2	M	10.010-Aspect Ration Matches Image Metadata	IF {10.011} EQ ASCII(JPEGB) OR ASCII(JPEGL) AND {10.008} NEQ 1 OR 2 THEN {10.009}/{10.010} EQ {HorizontalDensity-JPEGB,JPEGL} / {VerticalDensity-JPEGB,JPEGL} ELSE IF {10.011} EQ ASCII(JP2) OR ASCII(JP2L) <Provide Warning "Not Tested"> ELSE IF {10.011} EQ ASCII(PNG) AND {10.008} NEQ 1 OR 2 THEN {10.009}/{10.010} EQ { HorizontalDensity-PNG} / {VerticalDensity-PNG}	t-11	B
Field: 10.011-Compression Algorithm Value	8.10.11, Table 57, 7.7.9.3, 7.7.9.4	For non-facial images conveyed in Record Type-10 Field 10.011: Compression algorithm / CGA may be set to any value in Table 15, except WSQ20.		M	10.011-Value	<See Requirement ID: "Field: Type10 Compression">	t-2	
Field: 10.011-Compression Algorithm Metadata	8.10.11, Table 57	<The CGA is verified by checking the image metadata for the compression type signature if compression is used >	2	M	10.011-Matches Image Metadata	IF {10.011} EQ ASCII(JPEGB) OR ASCII(JPEGL) THEN Present(SOI - JPEG,JPEGL) AND IF {10.011} EQ ASCII(JP2) OR ASCII(JP2L) THEN Present(SigBox) AND IF {10.011} EQ ASCII(PNG) THEN Present(PNGSig)	t-11	B
Field: 10.012-Color Space Value	8.10.12, Table 57, 7.7.10	Table 16 lists the codes and their descriptions for each of the available color spaces used within this standard. All other color spaces are to be marked as undefined.		M	10.012-Value	<See Requirement ID: "Field: Image CSP Value>	t-2	
Field: 10.013-Subject Acquisition Profile Value	8.10.13, Table 57, 7.7.5.1	<Table 57 lists the value constraints for SAP.>		D	10.013-Value	<See Requirement ID: "Field: SAP Values">	t-2	
Field: 10.013-Subject Acquisition Profile Conditional	8.10.13, 7.7.5	The Subject Acquisition Profile (SAP) is a mandatory field when Field 10.003: Image type / IMT contains "FACE". Otherwise, it shall not be entered.		D	10.013-Conditional	<See Requirement ID: "Field: SAP Conditional">	t-2	
Field: 10.014-Face	8.10.14, Table 57	<Table 57 lists the value constraints for FIP.>	2	M	10.014-LHC-Value	{InfoItem:1 in 10.014} GTE 1 AND LTE {10.006} AND MO [integers]		B*

Field	Ref	Description	#	Req	Code	Constraint	Type
Image Bounding Box Value			2	M ⇐	10.014-RHC-Value	{InfoItem:2 in 10.014} GTE 1 AND LTE {10.006} AND MO [Integers]	B*
			2	M ⇐	10.014-TVC-Value	{InfoItem:3 in 10.014}) GTE 1 AND LTE {10.007} AND MO [Integers]	B*
			2	M ⇐	10.014-BVC-Value	{InfoItem:4 in 10.014}) GT {InfoItem:3 in 10.014} AND LTE {10.007} AND MO [Integers]	B*
			1	O ⇐	10.014-BBC-Value	{InfoItem:5 in 10.014}} MO [ASCII(S,H,F,N,X]	B*
Field: 10.014-Face Image Position Conditional	8.10.14	This field is only appropriate for images that do not comply with SAP Levels 30, 40, 50 or 51.		D	10.014-Conditional	<See Requirement ID: "Field: 10.014-FIP CondCode Dependent">	t-2
Field: 10.015-Face Image Path Value	8.10.15, Table 57	<Table 57 lists the value constraints for FPFI.>	1	M ⇐	10.015-BYC-Value	{InfoItem:1 in 10.015} MO [ASCII(C, E, P)]	B*
			1	M ⇐	10.015-NOP-Value	{InfoItem:2 in 10.015} MO [2 to 99] AND MO [Integers]	B*
			1	M ⇐	10.015-HPO-Value	For(X EQ 3 to {InfoItem:2 in 10.015}) { IF X MOD 2 EQ 1 {InfoItem:X in 10.015} GTE 0 AND LTE {10.006} AND MO [Integers] }	B*
			1	M ⇐	10.015-VPO-Value	For(X EQ 3 to {InfoItem:2 in 10.015}) { IF X MOD 2 EQ 0 {InfoItem:X in 10.015} GTE 0 AND LTE {10.007} AND MO [Integers] }	B*
Field: 10.016-Scanned Horizontal Pixel Scale Value	8.10.16, Table 57	<Table 57 lists the value constraints for SHPS.>		O	10.016-Value	<See Requirement ID: "Field: Image SHPS Value">	t-2
Field: 10.017-Scanned Verticall Pixel Scale Value	8.10.17, Table 57	<Table 57 lists the value constraints for SVPS.>		O	10.017-Value	<See Requirement ID: "Field: Image SVPS Value">	t-2

Field	Reference	Description	#	M/D	Field-ID	Condition / Constraint	Note	Char
Value								
Field: 10.018-Distortion Value	8.10.18, Table 57	<Table 57 lists the value constraints for DIST.>	1	M ⇐	10.018-IDK-Value	{InfoItem:1 in 10.018} MO [ASCII(Barrel, Inflated, Pincushion)]		B*
			1	M ⇐	10.018-IDM-Value	{InfoItem:2 in 10.018} EQ ASCII(E) OR ASCII(C)		B*
			1	M ⇐	10.018-DSC-Value	{InfoItem:3 in 10.018)} MO [ASCII(Mild, Moderate, Severe)]		B*
Field: 10.019-Lighting Artifacts Value	8.10.19, Table 57	<Table 57 lists the value constraints for LAF.>	1	D	10.019-LAF-Value	{10.019} MO [ASCII(F,H,R)]		B
Field: 10.020-Subject Pose Value	8.10.20, Table 60	When included, this field shall contain one character code selected from Table 60 to describe the pose of the subject.	1	D	10.020-Value	{10.020} MO [ASCII(F,R,LA,D)]		B
Field: YAW POA Opposite	8.10.20	Note that the offset angle in Field 10.021: Pose offset angle / POA is opposite from the yaw angle in Field 10.025 as indicated by a minus sign.	2	D	10.025-YAW, 10.021 Opposite	IF Present(InfoItem:1 in 10.025) THEN {10.021} EQ {InfoItem:1 in 10.025}*-1		B*
Field: 10.021-Pose Offset Angle Value	8.10.21, Table 57	When included, this field shall contain one character code selected from Table 60 to describe the pose of the subject.	1	D	10.021-Value	{10.021} GTE -180 AND LTE 180 AND {10.021} MO [Integers]		B
Field: 10.022-Deprecated	Table 57	Not to be used in new transactions.	-	-	10.022-Deprecated	<See Requirement ID: "Field: Type10-CondCode">.	t-2	
Field: 10.023-Photo Acquisition Source Value	8.10.22, Table 57, Table 57	When included, the first information item in this field shall contain an attribute code selected from Table 61 to describe the source of captured image data.	1	M ⇐	10.023-Value	{InfoItem:1 in 10.023} MO [ASCII(UNSPECIFIED, UNKOWN PHOTO, DIGITAL CAMERA, SCANNER, UNKNOWN VIDEO, ANALOG VIDEO, DIGITAL VIDEO, VENDOR, TYPE20, OTHER)]		B*
Field: 10.023-Photo Acquisition Source VENDOR	8.10.22, Table 57, Table 57	When "VENDOR" is specified in photo attribute code / PAC, a second free-format information item (vendor-specific description VSD) may be entered with up to 64 characters…	2	D	10.023-Dependent VSD	IF Present(InfoItem:2 in 10.023) THEN {InfoItem:1 in 10.023} EQ ASCII(VENDOR)	t-2	
Field: 10.023-Photo Acquisition Source Type-	8.10.22, Table 57, Table 57	A Record Type-20 may be used to store the original reference data. For this case, Field 10.997: Source representation / SOR shall be contained in this record, and the corresponding Record Type-20 shall be	2	D	10.023-Conditional	IF {InfoItem:1 in 10.023} EQ ASCII(TYPE20) THEN Present(10.997) AND Present(Record ST Type(Record) EQ 20)		B*

Field	Reference	Description	Count	M/O/D	ID	Constraint	Flag
20		included in the transaction.					
Field: 10.024-Subject Quality Scores Type-20	8.10.23, Table 57, 7.7.7	<Table 57 lists the value constraints for SQS.>		D	10.024-SQS	<See Requirement ID: "Field: Sample Quality Subfield 1" and "Field: Sample Quality Subfield 2" and "Field: Sample Quality Subfield 3".>	t-2
Field: 10.025-Subject Pose Angles Value	8.10.24, Table 57	<Table 57 lists the value constraints for SPA.>	1	M ⇑	10.025-YAW-Value	{InfoItem:1 in 10.025} GTE -180 AND LTE 180 AND MO [Integers]	B*
			1	M ⇑	10.025-PIT-Value	{InfoItem:2 in 10.025} GTE -90 AND LTE 90 AND MO [Integers]	B*
			1	M ⇑	10.025-ROL-Value	{InfoItem:3 in 10.025}} GTE -180 AND LTE 180 AND MO [Integers]	B*
			1	O ⇑	10.025-[YAWU, PITU, ROLU]-Value	{InfoItem:4 to 6 in 10.025} GTE 0 AND LTE 90 AND MO [Integers]	B*
Field: 10.026-Subject Facial Description Value	8.10.25, Table 57, Table 62	The value should be selected from the "Attribute code" column of Table 62.		D	10.026-SXS-Vlaue	TRUE <Cannot check for values because the standard allows user-defined Alphabetic Text. >	
Field: 10.027-Subject Eye Color Value	8.10.26, Table 57, Table 17	<Table 57 lists the value constraints for SEC.>		D	10.027-Value	<See Requirement ID: "Field: Image ECL Value">.	t-2
Field: 10.028-Subject Hair Color Value	8.10.27, Table 57, Table 63	<Table 57 lists the value constraints for SHC.> When the subject is predominantly bald, but hair color is discernible, then the appropriate hair color attribute code shall follow "BAL" in a second entry. For streaked hair, use "STR" in the first entry; use the second entry to describe the principal color of the hair.	1	M ⇑	10.028-SHC-Value	{InfoItem:1 in 10.028} MO [ASCII(XXX, BAL, BLK, BLN, BRO, GRY, RED, SDY, WHI, BLU, GRN, ONG, PNK, PLE, STR)] AND {InfoItem:2 in 10.028} MO [ASCII(XXX, BLK, BLN, BRO, GRY, RED, SDY, WHI, BLU, GRN, ONG, PNK, PLE)]	B*
Field: 10.029-2D Facial Feature Points Value	8.10.28, Table 57	<Table 57 lists the value constraints for FFP.> The first information item, feature point type / FPT is a one character value. It is	1	M ⇑	10.029-FPT-Value	ForEach(Subfield in 10.029) { {InfoItem:1 in Subfield} EQ 1 OR 2 }	B*
			2	M	10.029-	ForEach(Subfield in 10.029)	B*

Field	Reference	#	Char	Subfield	Description	Condition	Flag
			⇑	FPC-Value	mandatory. It shall be either 1 = Denoting an MPEG4 Feature point. 2 = Anthropometric landmark. (This is new to this version).	{ IF({InfoItem:1 in Subfield} EQ 1 THEN {InfoItem:2 in Subfield} MO [Figure 13, Figure 14] ELSE {InfoItem:2 in Subfield} MO [Table 65] }	B*
		1	M ⇑	10.029-HCX-Value	The second information item, feature point code / FPC is 3 to 5 characters. If FPTis 1, this information item shall be "A.B" with A and B defined in Section 8.10.27.1 and illustrated in Figure 14. The allowed special character is a period. If FPT is 2, the codes are entered as shown in the "Feature Point ID" column of Table 65. This is one to four alphabetic characters.	ForEach(Subfield in 10.029) { {InfoItem:3 in Subfield} GTE 1 AND LTE {10.006} AND MO [Integers] }	B*
		1	M ⇑	10.029-HCY-Value		ForEach(Subfield in 10.029) { {InfoItem:4 in Subfield} GTE 1 AND LTE {10.007} AND MO [Integers] }	
Field: 10.030-Device Monitoring Mode Value	8.10.29, Table 57,	1	O	10.030-Value	\<Table 57 lists the value constraints for DMM.\>	\<See Requirement ID: "Field: Device Monitoring"\>. t-2	
Field: 10.031-Tiered Markup Collection Value	8.10.30, Table 57,	1	D	10.031-Value	\<Table 57 lists the value constraints for TMC.\>	{10.031} MO [1 to 5, 100 to 999] AND MO [Integers]	B
Field: 10.032-3D Facial Feature Points Value	8.10.31, Table 57	1	M ⇑	10.032-FPT-Value	\<Table 57 lists the value constraints for 3DF.\> The first information item, feature point type / FPT is a one character value. It is mandatory. It shall be either 1 = Denoting an MPEG4 Feature point, but using a Z coordinate 2 = Anthropometric landmark, with a Z coordinate.	ForEach(Subfield in 10.032) { {InfoItem:1 in Subfield} EQ 1 OR 2 }	B*
		2	M ⇑	10.032-FPC-Value	The second information item, feature point code / FPC is 3 to 5 characters. If FPTis 1, this information item shall be	ForEach(Subfield in 10.032) { IF({InfoItem:1 in Subfield} EQ 1 THEN {InfoItem:2 in Subfield} MO [Figure 13, Figure 14] ELSE {InfoItem:2 in Subfield} MO [Table 65]	B*

"A.B" with A and B defined in Section 8.10.27.1 and illustrated in Figure 14. The allowed special character is a period. If FPT is 2, the codes are entered as shown in the "Feature Point ID" column of Table 65. Note that this entry is one to four alphabetic characters.						}		
		10.032-HCX-Value	M	⇐	1	ForEach(Subfield in 10.032) { [InfoItem:3 in Subfield] GTE 1 AND LTE {10.006} AND MO [Integers] }		B*
		10.032-HCY-Vlaue	M	⇐	1	ForEach(Subfield in 10.032) { [InfoItem:4 in Subfield] GTE 1 AND LTE {10.007} AND MO [Integers] }		B*
		10.032-HCZ-Value	M	⇐	1	ForEach(Subfield in 10.032) { [InfoItem:5 in Subfield] GTE 1 AND LTE 65535 AND MO [Integers] }		B*
Field: 10.033-Feature Contours Value	8.10.32, Table 57. Table 18	<Table 57 lists the value constraints for 3DF.>						
		10.033-FCC-Value	M	⇐	1	ForEach(Subfield in 10.033) { [InfoItem:1 in Subfield] MO [ASCII(eyetop, eyebottom, upperliptop, upperlipbottom, lowerliptop, lowerlipbottom, rightnostril, leftnostril, lefteyebrow, righteyebrow, chin, faceoutline] }		B*
		10.033-NOP-Value	M	⇐	1	ForEach(Subfield in 10.033) { [InfoItem:2 in Subfield] MO [3 to 99] AND MO [Integers] }		B*
		10.033-HPO-Value	M	⇐	1	ForEach(Subfield in 10.033) { For(X EQ 3 to [InfoItem:2 in Subfield]) IF X MOD 2 EQ 1 [InfoItem:X in Subfield] GTE 0 AND LTE {10.006} AND MO [Integers] } }		B*
		10.033-VPO-Value	M	⇐	1	ForEach(Subfield in 10.033) { For(X EQ 3 to [InfoItem:2 in Subfield]) {		B*

Field	Reference	Description	Condition	Value	Char	Occ	Type
Field: 10.034 to to 10.037-Reserved	Table 57	Reserved for future use only by ANSI/NIST-ITL.	IF X MOD 2 EQ 0 {InfoItem:X in Subfield} GTE 0 AND LTE {10.007} AND MO [Integers] } }	10.034 to 10.037-Reserved	-		
Field: 10.038-Comment Value	8.10.33, 7.4.4 Table 57	<Table 57 lists the value constraints for COM.>	<See Requirement ID: "Field: Type10-CondCode>.				t-2
Field: 10.039-Type-10 Reference Number Value	8.10.34, Table 57	<Table 57 lists the value constraints for T10.>	<See Requirement ID: "Field: Comment">.	10.038-Value	O	1	t-2
			{10.039} MO [1 to 255] AND MO [Integers]	10.039-Value	D	1	B
Field: 10.040-NCIC SMT Code Value	8.10.35, Table 57	<Table 57 lists the value constraints for SMT.>	ForEach(InfoItem in 10.040) {InfoItem} MO [NCIC Codes]	10.040-SMT-Value	D	1	B*
Field: 10.041-SMT Size Value	8.10.36, Table 57	<Table 57 lists the value constraints for SMS.>	ForEach(InfoItem in 10.041) {InfoItem} GTE 1 AND MO[Integers] }	10.041-[HGT, WID]	D	1	B*
Field: 10.042-SMT Descriptors Value	8.10.37, Table 57, Table 58, Table 67	<Table 57 lists the value constraints for SMD.>	ForEach(Subfield in 10.042) {InfoItem:1 in Subfield} MO [ASCII(SCAR,PIERCING, TATTOO, CHEMICAL, BRANDED, CUT, MARK)] }	10.042-SMI-Value	M ⇑	1	B*
			ForEach(Subfield in 10.042) {InfoItem:2 in Subfield} MO [ASCII(HUMAN, ANIMAL, PLANT, FLAG, OBJECT, ABSTRACT, SYMBOL, OTHER)] }	10.042-TAC-Value	D	1	B*
			ForEach(Subfield in 10.042) { IF [InfoItem:2 in Subfield} EQ ASCII(HUMAN) THEN [ASCII(MFACE,FFACE, ABFACE, MBODY, FBODY, ABBODY, ROLES, SPORT, MBPART, FBPART,	10.042-TSC-Value	D	2	B*

					ABBPART, MHUMAN, SKULL)]
					ELSE
					IF {InfoItem:2 in Subfield} EQ ASCII(ANIMAL)
					THEN
					{InfoItem:3 in Subfield} MO [ASCII(CAT, DOG, DOMESTIC, VICIOUS, HORSE, WILD, SNAKE, DRAGON, BIRD, INSECT, ABSTRACT, PARTS, MANIMAL)]
					ELSE
					IF {InfoItem:2 in Subfield} EQ ASCII(PLANT)
					THEN
					{InfoItem:3 in Subfield} MO [ASCII(NARCOTICS, REDFL, BLUEFL, YELFL, DRAW, ROSE, TULIP, LILY, MPLANT)]
					ELSE
					IF {InfoItem:2 in Subfield} EQ ASCII(FLAG)
					THEN
					{InfoItem:3 in Subfield} MO [ASCII(USA, STATE, NAZI, CONFED, BRIT, MFLAG)]
					ELSE
					IF {InfoItem:2 in Subfield} EQ ASCII(OBJECT)
					THEN
					{InfoItem:3 in Subfield} MO [ASCII(FIRE, WEAP,PLANE,VESSEL,TRAIN,VEHICLE,MYTH,SPOR T,NATURE, MOBJECTS)]
					ELSE
					IF {InfoItem:2 in Subfield} EQ ASCII(ABSTRACT)
					THEN
					{InfoItem:3 in Subfield} MO [ASCII(FIGURE, SLEEVE, BRACE, ANKLET, NECKLC, SHIRT, BODBND, HEDBND, MABSTRACT)]
					ELSE
					IF {InfoItem:2 in Subfield} EQ ASCII(SYMBOL)
					THEN
					{InfoItem:3 in Subfield} MO [ASCII(NATION, POLITIC, MILITARY, FRATERNAL, PROFESS, GANG, MSYMBOLS)]
					ELSE
					IF {InfoItem:2 in Subfield} EQ ASCII(OTHER)
					THEN
					{InfoItem:3 in Subfield} MO [ASCII(WORDING, FREEFRM, MISC)]
					}					

		1	D	10.042-[TAC, TSC]-Conditional	ForEach(Subfield in 10.042) { Present(InfoItem:2,3 in Subfield) IFF					B*

Field	Reference	Description	Count	M/D/O	Subfield	Rule	Level
						[InfoItem:1 in Subfield] NOT MO [ASCII[SCAR, MARK]]	B*
			1	D	10.042-TDS-Conditional	ForEach(Subfield in 10.042) { IF Present(InfoItem:4 in Subfield) , THEN [InfoItem:1 in Subfield] NOT MO [ASCII[SCAR, MARK]] }	B*
			1	D	10.042-TDS-Value Subfield 4	TRUE	
Field: 10.042 - SMT Descriptors Subfields Dependent	8.10.37, Table 57	...does not apply to scars and marks.		D	10.042-SubfieldCount Dependent	<See Requirement ID: "Field: Type10-Subfield Occurrence"> t-2	
Field: 10.043 - Tattoo Color Value	8.10.38, Table 57	<Table 57 lists the value constraints for COL.>	1	-	10.043-[TC1, TC2, TC3, TC4, TC5, TC6]-Value	ForEach(InfoItem in 10.043) { [InfoItem] MO [BLACK, BROWN, GRAY, BLUE, GREEN, ORANGE, PURPLE, RED, YELLOW, WHITE, MULTI, OUTLINE] }	B*
			2	D	10.043-SubfieldCount Matches 10.042	Count(Subfields in 10.043) EQ Count(Subfields in 10.042)	B*
Field: 10.044 - Image Transform Value	8.10.39, Table 57, Table 69	<Table 57 lists the value constraints for ITX.>	1	M ⇑	10.044-ITX-Value	ForEach(InfoItem in 10.044) { [InfoItem] MO [AGE, AXIS, COLORSHIFT, CONTRAST, CROP, DIST, DOWNSAMPLE, GRAY, ILLUM, IMGFUSE, INTERPOLATE, MULTCOMP, MULTIVIEW, POSE, ROTATE, SNIR, SUPERRES, WHITE] }	B*
Field: 10.045 - Occlusions Value	8.10.40, Table 20, Table 21	<Table 57 lists the value constraints for OCC.>	1	M ⇑	10.045-OCY-Value	ForEach(Subfield in 10.045) { [InfoItem:1 in Subfield] MO [ASCII[T, I, L, S]] }	B*
			1	M ⇑	10.045-OCT-Value	ForEach(Subfield in 10.045) { [InfoItem:2 in Subfield] MO [ASCII[H,S,C,R,O]] }	B*
			1	M ⇑	10.045-NOP-Value	ForEach(Subfield in 10.045) { [InfoItem:3 in Subfield] GTE 3 AND LTE 99 AND MO [Integers] }	B*

Field	Reference	Description	Count	M/O		Field ID	Condition/Content	Type		B
			1	M	⇑	10.045-HPO-Value	ForEach(Subfield in 10.045) { For(X EQ 1 to (InfoItem:3 in Subfield)) { IF X MOD 2 EQ 1 {InfoItem:X in Subfield} GTE 0 AND LTE {10.006} AND MO [Integers] } }			B*
			1	M	⇑	10.045-VPO-Value	ForEach(Subfield in 10.045) { For(X EQ 1 to (InfoItem:3 in Subfield)) { IF X MOD 2 EQ 0 {InfoItem:X in Subfield} GTE 0 AND LTE {10.007} AND MO [Integers] } }			B*
Field: 10.046 to 10.199-Reserved	Table 57	Reserved for future use only by ANSI/NIST-ITL.	-			10.046 to 10.199-Reserved	<See Requirement ID: "Field: Type10-CondCode">.	t-2		
Field: 10.200 to 10.900-User Defined	8.10.41, Table 57	User Defined Fields	-			10.200 to 10.900-User Defined	TRUE			B
Field: 10.901 Reserved	Table 57	Reserved for future use only by ANSI/NIST-ITL.	-			10.901-Reserved	<See Requirement ID: "Field: Type10-CondCode">.	t-2		
Field: 10.902-Annotated Information Value	8.10.42, Table 57	This is an optional field, listing the operations performed on the original source in order to prepare it for inclusion in a biometric record type. See Section 7.4.1.	O			10.902-[GMT, NAV, OWN, PRO]-Value	<See Requirement ID: "Field: xx.902-ANN">.	t-2		
Field: 10.903-Device Unique Identifier Value	8.10.43, Table 57	This is an optional field. See Section 7.7.1.1.	O			10.903-Value	<See Requirement ID: "Field: Device ID">.	t-2		
Field: 10.904-Make/Model/Serial Number	8.10.44, Table 57	This is an optional field. See Section 7.7.1.2.	O			10.904-[MAK, MOD, SER]-Value	<See Requirement ID: "Field: Make Model">.	t-2		

Field	Reference	Description			Field-Value	Requirement	Char Type		B
Value									
Field: 10.993- Source Agency Name	8.10.45, Table 57	This is an optional field. It may contain up to 125 Unicode characters.		O	10.993-Value	<See Requirment ID: "Field: Source Agency Name".>	t-2		
Field: 10.905 to 10.994-Reserved	Table 57	Reserved for future use only by ANSI/NIST-ITL.		-	10.905 to 10.994-Reserved	<See Requirement ID: "Field: Type10-CondCode">.	t-2		
Field: 10.995- Associated Context Value	8.10.46, Table 57	See Section 7 3.3		O	10.995-[ACN, ASP]-Value	<See Requirement IDs: "Field: xx.995-ASC" and "Field: xx.995-ASC-ACN" and "Field: xx.995-ASC-ASP">.	t-2		
Field: 10.996- Hash Value	8.10.47, Table 57	See Section 7 5.2		O	10.996-Value	<See Requirement ID: "Field: HAS">	t-2		
Field: 10.997- Source Representation Value	8.10.48, Table 57	See Section 7 3.2		O	10.997-[SRN, RSP]-Value	<See Requirement IDs: "Field: xx.997-SOR" and "Field: xx.997-SOR-SRN" and "Field: xx.997-SOR-RSP">.	t-2		
Field: 10.998- Geographic Sample Acquisition Location Value	8.10.49, Table 57	See Section 7.7.3		O	10.998-[UTE, LTD,LTM, LTS, LGD, LGM, LGS, ELE, GDC, GCM, GCE, GCN, GRT, OSI, OCV]-Value	<See Requirement IDs: "Field: Geographic", "Field: Geographic-Subfield 1" through "Field: Geographic-Values-SubField 15">.	t-2		
Field: 10.999- Image Data Valid	8.10.50, 7.7.9.4, Table 57	This is a mandatory field contains the image. For non-facial images conveyed in Record Type-10, Field 10.011: Compression algorithm/ CGA may be set to any value in Table 15, except WSQ20. <The image metadata is checked for validity.>	2	M	10.999-Uncompressed Image Length	IF {10.011} EQ ASCII(NONE) THEN Length(10.999) EQ 10.006)* {10.007}			B
			2	M	10.999-Valid Image Format	IF {10.011} EQ ASCII(JPEGB) OR ASCII(JPEGL) THEN Present(JFIF, SOI-JPEGL, SOF-JPEGB,JPEGL, EOI-JPEG, JPEGL) ELSE IF {10.011} EQ ASCII(JP2) OR ASCII(JP2L) THEN Present(SigBox, HeadBox,	t-11		B

```
ImgBox, EOI-JP2, JP2L)
ELSE
IF {10.011} EQ ASCII(PNG) THEN
    Present(PNGSig,
    IHDR,
    IDAT,
    IEND)
```

Table 6.10 - Assertions for Record Type 13 - Friction-Ridge Latent Image Record

Requirement ID	Reference in Base Standard	Requirement Summary	Lev el	S t a t u s	Assertion ID	Test Assertion	Test Note	Implementation Support	Supported Range	Test Result	Applicability
						8.13: Record Type-13: Friction-ridge latent image record					
Field: Type13-Subfield Occurrence	Table 70	<Table 70 specifies which fields contain subfields as well as the number of occurrences permitted >	1	M	13.[001 to 012, 016, 017, 020, 903, 904, 993, 996, 998, 999]-SubfieldCount	Count(Subfields in 13.[001 to 012, 016, 017, 020, 903,904, 993, 996, 998, 999]) EQ 1					T
			1	M	13.[001 to 012, 016, 017, 020, 903, 993, 996, 999]-InfoItemCount	Count(InfoItems in Subfield:1 in 13.[001 to 012, 016, 017, 020, 903, 993, 996, 999]) EQ 1					T
			1	M	13.013-SubfieldCount	Count(Subfields in 13.013) MO [1 to 6]					T
			1	M	13.013-InfoItemCount	ForEach(Subfield in 13.013) { Count(InfoItems in Subfield) EQ 1 }					T
			1	D	13.014-SubfieldCount	Count(Subfields in 13.014) MO [1 to 9]					T
			1	D	13.014-InfoItemCount	ForEach(Subfield in 13.014) { Count(InfoItems in Subfield) EQ 2 }					T
			1	D	13.015-SubfieldCount	<See Requirement ID: "Field: PPC-Subfield Occurrences">	t-2				T
			1	O	13.024-SubfieldCount	Count(Subfields in 13.024) MO [1 to 9]					T

1	O	13.024-InfoItemCount	ForEach(Subfield in 13.024) { Count(InfoItems in Subfield) EQ 4 }	T
1	O	13.902-SubfieldCount	Count (Subfields in 13.902) GTE 1	T
1	O	13.902-InfoItemCount	ForEach(Subfield in 13.902) { Count(InfoItems in Subfield) EQ 4 }	T
1	O	13.904-InfoItemCount	Count(InfoItems in 13.904) EQ 3	T
1	O	13.995-SubfieldCount	Count(Subfields in 13.995) MO [1 to 255]	T
1	O	13.995-InfoItemCount	ForEach(Subfield in 13.995) { Count(InfoItems in Subfield) EQ 1 OR 2 }	T
1	O	13.997-SubfieldCount	Count(Subfields in 13.997) MO [1 to 255]	T
1	O	13.997-InfoItemCount	ForEach(Subfield in 13.997) { Count(InfoItems in Subfield) EQ 1 OR 2 }	T
1	O	13.998-SubfieldCount	<See Requirement ID: "Field: Geographic"> t-2	B
1	-	[13.001 to 13.013, 13.999]-Mandatory CondCode	Present(13.001 to 13.013, 13.999)	B
1	-	[13.018, 13.019, 13.021 to 13.023, 13.025 to 13.199, 13.901, 13.905 to 13.992, 13.994]-	NOT Present(13.018, 13.019, 13.021 to 13.023, 13.025 to 13.199, 13.901, 13.905 to 13.992, 13.994)	B

Table 70   <Table 70 specifies the Condition Code for each field.>

Field: Type13-CondCode

Field	Reference	Description		Element	Requirement / Condition		Char Type
				Reserved			
Field: 13.014-Search Position Descriptors Dependent	Table 70, 8.13.14	This field shall be present if and only if the finger position code "19" appears in Field 13.013: Friction ridge generalized position / FGP.	D	13.014-CondCode Dependent	<See Requirement ID: "Field: SPD PPC Conditional">	t-2	B
Field: 13.015-Print Position Coordinates Dependent	Table 57, 8.13.15	This field may be present if and only if the finger position code "19" appears in Field 13.013: Friction ridge generalized position / FGP.	D	13.015-CondCode Dependent	<See Requirement ID: "Field: SPD,PPC Conditional">	t-2	
Field: Type13-CharType	8.13, Table 70	<Table 70 specifies the Character Type for each field that contains no subfields. Note that even though some Character Types are listed as U (user defined), they may still have character type limitations defined in the standard text. >	-	13.[001,002,003, 006 to 010, 012, 016, 017]-CharType	Bytes(13.[001,002,003, 006 to 010, 012, 016, 017]) MO [0x30 to 0x39]		B
			-	13.011-CharType	Bytes(13.011) MO [0x20, 0x41 to 0x5A, 0x61 to 0x7A, 0x30 to 0x39]		B
			-	[13.004, 13.020, 13.993]-CharType	Present(Bytes(13.004, 13.020, 13.993))		B
			M	13.005-CharType	Bytes(13.005) MO [0x30 to 0x39]		T
			M	NIEM-13.005-CharType	Bytes(13.005) MO [0x30 to 0x39, 0x2D]		X
			O	13.903-CharType	Bytes(13 903) MO [0x20 to 0x7E]		B
			O	13.996-CharType	Bytes(13 996) MO [0x30 to 0x39, 0x41 to 0x46, 0x61 to 0x66]		B
			M	13.999-CharType	TRUE		B
Field: Type13-Subfield CharType	8.13, Table 70	<Table 70 specifies the Character Type for each subfield.>	M	13.013-FGP-CharType	Bytes(All(InfoItems in 13.013)) MO [0x30 to 0x39]		B*
			D	13.014-PDF-CharType	ForEach(Subfield in 13.014) { Bytes(InfoItem:1 in Subfield) MO [0x30 to 0x39] }		B*
			D	13.014-FIC-CharType	ForEach(Subfield in 13.014) { Bytes(InfoItem:2 in Subfield) MO [0x30 to 0x39, 0x20, 0x41 to 0x5A, 0x61 to 0x7A] }		B*

1	D	13.015-FVC-CharType	ForEach(Subfield in 13.015) { Bytes(InfoItem:1 in Subfield)) MO [0x30 to 0x39, 0x20, 0x41 to 0x5A, 0x61 to 0x7A] }	B*
1	D	13.015-LOS-CharType	ForEach(Subfield in 13.015) { Bytes(InfoItem:2 in Subfield)) MO [0x20, 0x41 to 0x5A, 0x61 to 0x7A] }	B*
1	D	13.015-[LHC, RHC, TVC, BVC]-CharType	ForEach(Subfield in 13.015) { Bytes(InfoItem:3 to 6 in Subfield)) MO [0x30 to 0x39] }	B*
1	O	13.024-[FRMP, QVU, QAP]-CharType	ForEach(Subfield in 13.024) { Bytes(InfoItem:1,2,4 in Subfield)) MO [0x30 to 0x39] }	B*
1	O	13.024-QAV-CharType	ForEach(Subfield in 13.024) { Bytes(InfoItem:3 in Subfield)) MO [0x30 to 0x39,0x41 to 0x46, 0x61 to 0x66] }	B*
1	O	13.902-[NAV, OWN, PRO]-CharType	TRUE	T
1	O	13.902-GMT-CharType	ForEach(Subfield in 13.902) { Bytes(InfoItem:1 in Subfield) MO [0x30 to 0x39, 0x5A] }	T
1	O	NIEM-13.902-Subfield	< The treatment of subfields for validation in the XML version requires further review. Byte values allowed for first "subfield" in XML are 0x30 to 0x39, 0x3A, 0x54, 0x5A.>	X*
1	O	13.904-[MAK, MOD, SER]-CharType	TRUE	B*
1	O	13.995-[CAN, ASP]-CharType	Bytes(All(InfoItem:1,2 in 13 995)) MO [0x30 to 0x39]	B*
1	O	13.997-	Bytes(All(InfoItem:1,2 in 13 997)) MO [0x30 to	B*

Field: Type13-CharCount							
		[SRN, RSP]-CharType	0x39		O		
		13.998-[UTE, LTD,LTM, LTS, LGD, LGM, LGS, ELE, GDC, GCM, GCE, GCN, GRT, OSI, OCV]-CharType	<See Requirement ID: "Field: Geographic">	t-2			T
	Table 70	13.001-CharCount	DataLength(13.001) MO [1 to 8]		M	1	T
	<Table 70 specifies the Character Count for each field that contains no subfields.>	NIEM-13.001-CharCount	Length(13.001) EQ 2		M	1	X
		13.002-CharCount	DataLength(13.002) EQ 1 OR 2		M	1	B
		13.003-CharCount	DataLength(13.003) EQ 1 OR 2		M	1	B
		13.004-CharCount	<See Requirement ID: "Field: Originating Agency".>	t-2			T
		13.005-CharCount	DataLength(13.005) EQ 8		M	1	
		NIEM-13.005-CharCount	DataLength(13.005) EQ 10		M	1	X
		13.006-CharCount	DataLength(13.006) MO [2 to 5]		M	1	B
		13.007-CharCount	DataLength(13.007) MO [2 to 5]		M	1	B
		13.008-CharCount	DataLength(13.008) EQ 1		M	1	B
		13.009-CharCount	DataLength(13.009) MO [1 to 5]		M	1	B
		13.010-CharCount	DataLength(13.010) MO [1 to 5]		M	1	B
		13.011-CharCount	DataLength(13.011) MO [3 to 5]		M	1	B
		13.012-CharCount	DataLength(13.012) EQ 1 OR 2		M	1	B
		13.016-CharCount	DataLength(13.016) MO [1 to 5]		O	1	B

Field	Table/Description	Occ	M/O/D	Subfield	Condition	B
		1	O	13.017-CharCount	DataLength(13.017) MO [1 to 5]	B
		1	O	13.020-CharCount	DataLength(13.020) MO [1 to 126]	B
		1	O	13.903-CharCount	DataLength(13.903) MO [13 to 16]	B
		1	O	13.993-CharCount	<See Requirment ID: "Field: Source Agency Name".> t-2	B
		1	O	13.996-CharCount	DataLength(13.995) EQ 64	B
		1	O	13.999-CharCount	DataLength(13.999) GTE 1	B
Field: Type13-Subfield CharCount	Table 70 <Table 70 specifies the Character Count for each subfield.>	1	M	13.013-FGP-CharCount	Length(All(InfoItems in 13.013)) EQ 1 OR 2	B*
		1	D	13.014-PDF-CharCount	ForEach(Subfield in 13.014) { Length(InfoItem:1 in Subfield) EQ 1 OR 2 }	B*
		1	D	13.014-FIC-CharCount	ForEach(Subfield in 13.014) { Length(InfoItem:2 in Subfield) EQ 3 }	B*
		1	D	13.015-[FVC, LOS]-CharCount	ForEach(Subfield in 13.015) { Length(InfoItem:1,2 in Subfield)) EQ 2 OR 3 }	B*
		1	D	13.015-[LHC, RHC, TVC, BVC]-CharCount	ForEach(Subfield in 13.015) { Length(InfoItem:3 to 6 in Subfield)) MO [1 to 5] }	B*
		1	O	13.024-FRMP-CharCount	ForEach(Subfield in 13.024) { Length(InfoItem:1 in Subfield)) EQ 1 OR 2 }	B*
		1	O	13.024-QVU-CharCount	ForEach(Subfield in 13.024) { Length(InfoItem:2 in Subfield)) MO [1 to 3] }	B*
		1	O	13.024-CharCount	ForEach(Subfield in 13.024)	B*

		Field	Definition	
1		QAV-CharCount	{ Length(InfoItem:3 in Subfield)) EQ 4 }	B*
1	O	13.024-QAP-CharCount	ForEach(Subfield in 13.024) { Length(InfoItem:4 in Subfield)) MO [1 to 5] }	T
1	O	13.902-GMT-CharCount	ForEach(Subfield in 13.902) { Length(InfoItem:1 in Subfield) EQ 15 }	T
1	O	13.902[NAV, OWN]-CharCount	ForEach(Subfield in 13.902) { Length(InfoItem:2,3 in Subfield) MO [1 to 64] }	T
1	O	13.902-PRO-CharCount	ForEach(Subfield in 13.902) { Length(InfoItem:4 in Subfield)) MO [1 to 255] }	X*
1	O	NIEM-13.902-Subfield-CharCount	< The treatment of subfields for validation in the XML version requires further review. Length of the first "subfield" in XML is 20.>	B*
1	O	13.904–[MAK, MOD, SER]-CharCount	Length(All((InfoItems in 13.904)) MO [1 to 50]	B*
1	O	13.995-ACN-CharCount	ForEach(Subfield in 13.995) { Length(InfoItem:1 in Subfield) MO [1 to 3] }	B*
1	O	13.995-ASP-CharCount	ForEach(Subfield in 13.995) { Length(InfoItem:2 in Subfield) EQ 1 OR 2 }	B*
1	O	13.997-SRN-CharCount	ForEach(Subfield in 13.997) { Length(InfoItem:1 in Subfield) MO [1 to 3] }	B*
1	O	13.997-RSP-CharCount	ForEach(Subfield in 13.997) { Length(InfoItem:2 in Subfield) MO [1 to 2] }	B*

114

Field	Reference	Description			Field ID	Requirement	Type						Char
Field: Type13-Field Occurrence	Table 70	<Table 70 specifies the Field Occurrence for each field.>	1	O	13.998-[UTE, LTD,LTM, LTS, LGD, LGM, LGS, ELE, GDC, GCM, GCE, GCN, GRT, OSJ, OCV]-CharCount	<See Requirement ID: "Field: Geographic">	t-2					B
			1	-	13.[018, 019, 021 to 023, 025 to 199,901,905 to 992, 994]-Occurrence	Count(13.[018, 019, 021 to 023, 025 to 199,901,905 to 992, 994]) EQ 0							B
			1	M	13.[001 to 013, 999]-Occurrence	Count(13.[001 to 013, 999]) EQ 1							B
			1	-	13.[014 to 017,020, 024, 902 to 904, 993, 995 to 998]-Occurrence	Count(13.[014 to 017,020, 024, 902 to 904, 993, 995 to 998]) LTE 1							
Field: 13.001-Record Header Value	8.13.1, Table 70, 7.1	Field 13.001 Record header. In Traditional encoding, this field contains the record length in bytes (including all information separators)	1	M	13.001-Record Header	<See Requirement ID "Field: xx.001-Record Header">	t-2						
	8.13.1, C.9.11	The XML name for the Type-13 record is <itl:PackageLatentImageRecord>, and its <biom:RecordCategoryCode> element shall have a value of "13".	1	M	NIEM-13.001-Value	ForEach(itl:PackageLatentImageRecord){ (XElm(biom::RecordCategoryCode)} EQ ASCII(13) }							X
Field: 13.002-Information Designation Character Value	8.13.2, Table 70, 7.3.1	This mandatory field shall contain the IDC assigned to this Type-13 record as listed in the information item IDC for this record in Field 1.003 Transaction content/CNT.	1	M	13.002-IDC	<See Requirement IDs "Field: xx.002-IDC and "Field: 1.003-Transaction Content Subfield 2 IDC Matches" >	t-2						
Field: 13.003-Impression Type Value	8.13.3, Table 70, 7.7.4.1	This mandatory field shall indicate the manner by which the latent print was obtained. See Section 7.7.4.1 for details. Valid values are 4 through 7, 12 through 15, 28 or 29, and 32 through 39.	1	M	13.003-Value	{13.003}MO [4 to 7, 12 to 15, 28, 29, 32 to 39] AND MO [integers]							B

Field	Reference	Description	#	M	Value ID	Requirement	t-code	Type
Field: 13.004-Source Agency Value	8.13.4, 7.6	This is a mandatory field. See Section 7.6 for details.		M	13.004-Value	<See Requirement ID: "Field: Originating Agency".>	t-2	
Field: 13.005-Latent Capture Date Value	8.13.5, 7.7.2.3	This mandatory field shall contain the date that the latent biometric data contained in the record was captured.	1	M	13.005-Value	{13.005} MO [ValidLocalDate]	t-6	T
			1	M	13.005-NIEM-13.005-Value	ForEach(XElm(itl:PackageLatentImageRecord)){ XElm(nc:Date) in XElm(biom:CaptureDate)} MO [NIEM-ValidLocalDate] }	t-6	X
Field: 13.006-Horizontal Line Length Value	8.13.6, Table 70, 7.7.8.1	The maximum horizontal size is limited to 65,534 pixels in Record Types-4 and 8, and to 99,999 for other record types. The minimum value is 10 pixels.	2	M	13.006-Value	<See Requirement ID "Field: Image HLL Value" >	t-2	
Field: 13.006-Horizontal Line Length Metadata	8.13.6, Table 70, 7.7.8.1	<The HLL is verified by checking the image metadata if compression is used.>	2	M	13.006-Matches Image Metadata	IF {13.011} EQ ASCII(JPEGL) THEN {13.006} EQ {ImageWidth-JPEGB,JPEGL} ELSE IF {13.011} EQ ASCII(JP2L) THEN {13.006} EQ {ImageWidth-JP2,JP2L} ELSE IF {13.011} EQ ASCII(PNG) THEN {13.006} EQ {ImageWidth-PNG} ELSE IF {13.011} EQ ASCII(WSQ20) THEN {13.006} EQ {ImageWidth-WSQ}	t-11	B
Field: 13.007-Vertical Line Length Value	8.13.7, Table 70, 7.7.8.2	The maximum vertical size is limited to 65,534 pixels in Record Types-4 and 8, and to 99,999 for other record types. The minimum value is 10 pixels.	2	M	13.007-Value	<See Requirement ID "Field: Image VLL Value">	t-2	
Field: 13.007-Vertical Line Length Metadata	8.13.7, Table 70, 7.7.8.2	<The VLL is verified by checking the image metadata if compression is used.>	2	M	13.007-VLL Metadata JPEGL	IF {13.011} EQ ASCII (JPEGL) THEN {13.007} EQ {ImageHeight-JPEGB,JPEGL} ELSE IF {13.011} EQ ASCII(JP2L) THEN {13.007} EQ {ImageHeight-JP2,JP2L} ELSE IF {13.011} EQ ASCII(PNG) THEN {13.007} EQ {ImageHeight-PNG} ELSE IF {13.011} EQ ASCII(WSQ20) THEN {13.007} EQ {ImageHeight-WSQ}	t-11	B
Field: 13.008-Scale	8.13.8, Table 70,	<Table 70 lists the value constraints for SLC>		M	13.008-Value	<See Requirement ID "Field: Image SLC Value" >	t-2	

Field	Ref	Description			Requirement	Logic	Test	
Units Value	7.7.8.3							
Field: 13.008-Scale Units Metadata	8.13.8, Table 70, 7.7.8.3	A value of "1" shall indicate pixels per inch. A value of "2" shall indicate pixels per centimeter. A value of "0" in this field indicates that no scale is provided, and the quotient of THPS/TVPS shall provide the pixel aspect ratio. <The SLC is verified by checking the image metadata if compression is used.>	2	M	13.008-Matches Image Metadata	IF {13.011} EQ ASCII(JPEGL) THEN {13.008} EQ {SamplingUnits-JPEGB,JPEGL} ELSE IF ASCII(JP2L) OR ASCII(WSQ20) THEN <Provide Warning "Not Tested"> ELSE IF {13.011} EQ ASCII(PNG) THEN IF {13.008} EQ 1 OR 2 THEN {SamplingUnits-PNG} EQ 1, ELSE IF {13.008} EQ 0 THEN {SamplingUnits-PNG} EQ 0	t-11	B
Field: 13.009-Transmitted Horizontal Pixel Scale Value	8.13.9, Table 70, 7.7.8.4	<Table 70 lists the value constraints for THPS.>		M	13.009-Value	<See Requirement ID "Field: Image THPS Value">	t-2	
Field: 13.009-Transmitted Horizontal Pixel Scale Metadata	8.13.9, Table 70, 7.7.8.4	This is the integer pixel density used in the horizontal direction of the image if SLC has a value of "1" or "2". If SLC has a value of "0", this information item shall contain the horizontal component of the pixel aspect ratio, up to 5 digits. <The THPS is verified by checking the image metadata if compression is used >	2	M	13.009-Matches Image Metadata	IF {13.011} EQ ASCII(JPEGL) AND {13.008} EQ 1 OR 2 THEN {13.009} EQ {HorizontalDensity-JPEGB,JPEGL} ELSE IF ASCII(JP2L) OR ASCII(WSQ20) THEN <Provide Warning "Not Tested"> ELSE IF {13.011} EQ ASCII(PNG) AND {13.008} EQ 1 THEN {13.009} EQ {HorizontalDensity-PNG} * 0.0254 (meters/inch), ELSE IF {13.011} EQ ASCII(PNG) AND {13.008} EQ 2 THEN {13.009} EQ {HorizontalDensity-PNG} * 0.01 (meters/cm)	t-11, t-12	B
			2	M	13.009-Aspect Ratio Matches Image Metadata	IF {13.011} EQ ASCII(JPEGL) AND {13.008} NEQ 1 OR 2 THEN {13.009}/{13.010} EQ {HorizontalDensity-JPEGB,JPEGL} / {VerticalDensity-JPEGB,JPEGL} ELSE IF ASCII(JP2L) OR ASCII(WSQ20) THEN <Provide Warning "Not Tested"> ELSE IF {13.011} EQ ASCII(PNG) AND {13.008} NEQ 1 OR 2 THEN {13.009}/{13.010} EQ {HorizontalDensity-PNG} / {VerticalDensity-PNG}	t-11	B
Field: 13.010-Transmitted	8.13.10, Table 70, 7.7.8.5	<Table 70 lists the value constraints for TVPS.>		M	13.010-value	<See Requirement ID "Field: Image TVPS Value">	t-2	

Field	Reference	Description		M	Check	Validation Logic	Test						B
Vertical Pixel Scale Value													
Field: 13.010-Transmitted Vertical Pixel Scale Metadata	8.13.10, Table 70, 7.7.8.5	This is the integer pixel density used in the Vertical direction of the image if SLC has a value of "1" or "2". If SLC has a value of "0", this information item shall contain the Vertical component of the pixel aspect ratio, up to 5 digits. \<The TVPS is verified by checking the image metadata if compression is used.>	2	M	13.010-Mathces Image Metadata	IF {13.011} EQ ASCII(JPEGL) AND {13.008} EQ 1 OR 2 THEN {13.010} EQ {VerticalDensity-JPEGB,JPEGL} ELSE IF ASCII(JP2L) OR ASCII(WSQ20) THEN \<Provide Warning "Not Tested"> ELSE IF {13.011} EQ ASCII(PNG) AND {13.008} EQ 1 THEN {13.010} EQ {VerticalDensity-PNG} * 0.0254 (meters/inch), ELSE IF {13.011} EQ ASCII(PNG) AND {13.008} EQ 2 THEN {13.010} EQ {VerticalDensity-PNG} * 0.01 (meters/cm)	t-11, t-12						B
			2	M	13.010-Aspect Ratio Mathces Image Metadata	IF {13.011} EQ ASCII(JPEGL) AND {13.008} NEQ 1 OR 2 THEN {13.009}/{13.010} EQ {HorizontalDensity-JPEGB,JPEGL} / {VerticalDensity-JPEGB,JPEGL} ELSE IF ASCII(JP2L) OR ASCII(WSQ20) THEN \<Provide Warning "Not Tested"> ELSE IF {13.011} EQ ASCII(PNG) AND {13.008} NEQ 1 OR 2 THEN {13.009}/{13.010} EQ { HorizontalDensity-PNG}/{VerticalDensity-PNG}	t-11						B
Field: 13.011-Compression Algorithm Value	8.13.11, Table 70, 7.7.9.1, 5.3.13	For each of these fields, the entry corresponds to the appropriate *Label* entry in Table 15: Field 13.011: Compression algorithm / CGA. The variable-resolution for latent image data contained in the Type-13 record shall be uncompressed or may be the output from a lossless compression algorithm.	1	M	13.011-Value	{13.011} MO [ASCII(NONE, JPEGL, JP2L, PNG, WSQ)]							B
Field: 13.011-Compression Algorithm Metadata	8.13.11, Table 70	\<The CGA is verified by checking the image metadata for the compression type signature if compression is used.>	2	M	13.011-Matches Image Metadata	IF {13.011} EQ ASCII(JPEGL) THEN Present(SOI -JPEGJPEGL) ELSE IF {13.011} EQ ASCII(JP2L) THEN Present(SigBox) ELSE IF {13.011} EQ ASCII(PNG) THEN Present(PNGSig) ELSE	t-11						B

Field	Reference	#	M/D	Req ID	Description	Condition	Test	
						IF {13.011} EQ ASCII(WSQ20) THEN Present(SOI-WSQ)		B
Field: 13.012-Bits Per Pixel Value	8.13.12, Table 70, 7.7.8.6		M	13.012-Value	This field shall contain an entry of "8" for normal grayscale values of "0" to "255". Any entry in this field greater than "8" shall represent a grayscale pixel with increased proportion.	IF {13.011} EQ ASCII(WSQ20) THEN Present(SOI-WSQ) <See Requirement ID "Field: Image BPX Value" >	t-2	
Field: 13.012- Bits Per Pixel Metadata	8.13.12, Table 70	2	M	13.012-Matches Image Metadata	<The BPX is verified by checking the image metadata for the compression type signature if compression is used>	IF {13.011} EQ ASCII(JPEGL) THEN {13.012} EQ {BPX-JPEG, JPEGL} ELSE IF {13.011} EQ ASCII(JP2L) THEN {13.012} EQ {BPX-JP2, JP2L} ELSE IF {13.011} EQ ASCII(PNG) THEN {13.012} EQ {BPX-PNG} ELSE IF {13.011} EQ ASCII(WSQ20) THEN <Provide Warning "Not Tested">	t-11	B
Field: 13.013- Friction Ridge Generalized Position Value	8.13.13, Table 70, 7.7.4.2, Table 8	1	M	13.013-FGP-Value	See Section 7.7.4.2 and Table 8 for details.	{All(InfoItems in 13.013)} MO [0 to 38, 40 to 50, 60 to 79, 81 to 84] AND MO [Integers]	t-2	
Field: 13.013- Friction Ridge Generalized Position Conditional	8.13.13, Table 70		M	13.013-FGP-Conditional	If code 19 is used, fields 13.014 and 13.015 shall be used.	<See Requirement ID: "Field: SPD,PPC Conditional">	t-2	
Field: 13.014- Search Position Descriptors Value	8.13.14, Table 70, 7.7.4.3		D	13.014-[PDF, FIC] Value	...described in Section 7.7.4.3	<See Requirement ID: "Field: SPD PPD Value">	t-2	
Field: 13.014- Search Position Descriptors Conditional	8.13.14, Table 70		D	13.014-Conditional	This field shall be present if and only if the finger position code "19" appears in Field 13.013: Friction ridge generalized position / FGP.	<See Requirement ID: "Field: SPD,PPC Conditional">	t-2	
Field: 13.015-Print	8.13.15, Table 70,		D	13.015-[LOS, LHC,	See section 7.7.4.4	<See Requirement IDs: "Field: PPC-Subfield 1" through "Field: PPC-SubfieldCount 5 6".>	t-2	

Field	Section	Description	Code	Field ID	Requirement / Value	Type
Position Coordinates Value	7.7.4.4			RHC, TVC, BVC]-Value		
Field: 13.015- Print Position Coordinates Conditional	8.13.15, Table 70	This field may be present if and only if the finger position code "19" appears in Field 13.013: Friction ridge generalized position / FGP.	D	13.015- Conditional	<See Requirement ID: "Field: SPD PPC Conditional">	t-2
Field: 13.016- Scanned Horizontal Pixel Scale Value	8.13.16, Table 70, 7.7.8.7	See section 7.7.8.7 for details.	O	13.016- Value	<See Requirement IDs: "Field: Image SHPS Value">	t-2
Field: 13.017- Scanned Vertical Pixel Scale Value	8.13.17, Table 70, 7.7.8.8	See section 7.7.8.8 for details.	O	13.017- Value	<See Requirement IDs: "Field: Image SVPS Value">	t-2
Field: 13.018, 13.019- Reserved	Table 70	Reserved for future use only by ANSI/NIST-ITL.	-	[13.018, 13.019]-Reserved	<See Requirement ID: "Field: Type13-CondCode>.	t-2
Field: 13.020- Comment Value	8.13.17, Table 70, 7.4.4	See section 7.4.4 for details.	O	13.020- Value	<See Requirement ID: "Field: Comment>.	t-2
Field: 13.021, 13.023- Reserved	Table 70	Reserved for future use only by ANSI/NIST-ITL.	-	[13.021, 13.023]-Reserved	<See Requirement ID: "Field: Type13-CondCode>.	t-2
Field: 13.024- Latent Quality Metric Value	Table 70, Table 8, 8.13.19	The first information item is the entry in Field 13.013: Friction ridge generalized position / FGP, as chosen from Table 8.... See Section 7.7.7 for a description of the remaining three information items.	O	13.024- [FRMP, QVU, QAV, QAP]-Value	<See Requirement IDs: "Field: Sample Quality Subfield 1", "Field: Sample Quality Subfield 2", "Field: Sample Quality Subfield 3", "Field: Sample Quality Additional Subfield">.	t-2
Field: 13.025 to 13.199- Reserved	Table 70	Reserved for future use only by ANSI/NIST-ITL.	-	13.025, 13.199-Reserved	<See Requirement ID: "Field: Type13-CondCode>.	t-2
Field: 13.200 to 13.900- User Defined	Table 70	User Defined Fields	-	13.200 to 13.900- User Defined	TRUE	B
Field:	Table 70	Reserved for future use only by	-	13.901-	<See Requirement ID: "Field: Type13-	t-2

Field	Section	Description	Cond.	Value	Requirement ID	Type
13.901-Reserved		ANSI/NIST-ITL.		Reserved	CondCode>.	
Field: 13.902-Annotated Information Value	8.13.21, Table 70	This is an optional field, listing the operations performed on the original source in order to prepare it for inclusion in a biometric record type. See Section 7.4.1.	O	13.902-[GMT, NAV, OWN, PROJ]-Value	<See Requirement ID: "Field: xx.902-ANN" >.	t-2
Field: 13.903-Device Unique Identifier Value	8.13.22, Table 70	This is an optional field. See Section 7.7.1.1.	O	13.903-Value	<See Requirement ID: "Field: Device ID" >.	t-2
Field: 13.904-Make/Model/Serial Number Value	8.13.23, Table 70	This is an optional field. See Section 7.7.1.2.	O	13.904-[MAK, MOD, SER]-Value	<See Requirement ID: "Field: Make Model" >.	t-2
Field: 13.905 to 13.992-Reserved	Table 70	Reserved for future use only by ANSI/NIST-ITL.	-	13.905, 13.994-Reserved	<See Requirement ID: Type13-CondCode>.	t-2
Field: 13.993-Source Agency Name	8.13.24, Table 70	This is an optional field. It may contain up to 125 Unicode characters.	O	13.993-Value	<See Requirement ID: "Field: Source Agency Name".>	t-2
Field: 13.994-Reserved	Table 70	Reserved for future use only by ANSI/NIST-ITL.	-	13.994-Reserved	<See Requirement ID: "Field: Type13-CondCode>.	t-2
Field: 13.995-Associated Context Value	8.13.24, Table 70	See Section 7.3.3	O	13.995-[ACN, ASP]-Value	<See Requirement IDs: "Field: xx.995-ASC" and "Field: xx.995-ASC-ACN" and "Field: xx.995-ASC-ASP">.	t-2
Field: 13.996-Hash Value	8.13.25, Table 70	See Section 7.5.2	O	13.996-Value	<See Requirement ID: "Field: HAS">	t-2
Field: 13.997-Source Representation Value	8.13.26, Table 70	See Section 7.3.2	O	13.997-[SRN, RSP]-Value	<See Requirement IDs: "Field: xx.997-SOR-SRN" and "Field: xx.997-SOR-RSP">.	t-2
Field: 13.998-Geographic	8.13.27, Table 70	See Section 7.7.3	O	13.998-[UTE, LTM, LTD,LTM,	<See Requirement IDs: "Field: Geographic", "Field: Geographic", "Field: Geographic-Subfield 1" through "Field: Geographic-Values-SubField	t-2

							B
							B
					t-11, t-16		B

Sample Acquisition Location Value			LTS, LGD, LGM, LGS, ELE, GDC, GCM, GCE, GCN, GRT, OSI, OCV]-Value	15" >.		
Field: 13.999- Image Data Valid	8.13.28, Table 70	This is a mandatory field contains the image.	2	M	13.999- Uncompressed Image Length	IF {13.011} EQ ASCII(NONE) THEN Length(13.999) EQ 13.006} * {13.007}
		<The image metadata is checked for validity.>	2	M	13.999- Valid Image Format	IF {13.011} EQ ASCII(NONE) THEN Length(13.999) EQ 13.006} * {13.007} ELSE IF {13.011} EQ ASCII(JPEGL) THEN Present(JFIF, SOI-JPEGB, JPEGL, SOF-JPEGB, JPEGL, EOI-JPEG, JPEGL) ELSE IF {13.011} EQ ASCII(JP2L) THEN Present(SigBox, HeadBox, ImgBox, EOI-JP2L) ELSE IF {13.011} EQ ASCII(PNG) THEN Present(PNGSig, IHDR, IDAT, IEND) ELSE IF {13.999} EQ ASCII(WSQ20) THEN Present(SOI-WSQ,SOF-WSQ,SOB-WSQ,EOI-WSQ)
Field: 13.999- Image WSQ Version 3.1	7.7.9.1	Only version 3.1 or higher shall be used for compressing grayscale fingerprintdata at 500 ppi class with a platen area of 2 inches or greater in height. WSQ 2.0 or higher may be used for 500 ppi class data taken from a platen of less than 2 inches in height. WSQ shall not be used for other than the 500 ppi class.	2	M	13.999- Valid WSQ Encoder Version	IF {13.011} EQ ASCII(WSQ20) THEN {Encoder Version} EQ 1 OR 2

Table 6.11 - Assertions for Record Type 14 - Fingreprint Image Record

122

Requirement ID	Reference in Base Standard	Requirement Summary	Lev	Stat	Assertion ID	Test Assertion	Test Note	Implementation Support	Supported Range	Test Result	Applicability
						8.14: Record Type-14: Fingerprint image record					
Record: Type14-Fingerprint Type	8.14	The Type-14 record shall contain and be used to exchange exemplar fingerprint image data, such as a rolled tenprint, an identification flat, or a complete friction ridge exemplar. All fingerprint impressions shall be acquired from a card, a single or multiple-finger flat-capture device, contactless fingerprint sensor that outputs 2D fingerprint images, or a live-scan device. Captured images may be transmitted to agencies that will automatically extract the desired feature information from the images for matching purposes.	3	M	Type14-Fingerprint Type	<Unsupported: It is not feasible to test if the image represents an exemplar fingerprint or how the image was acquired.>	t-1				
Field: Type14-Subfield Occurrence	Table 71	<Table 71 specifies which fields contain subfields as well as the number of occurrences permitted >	1	M	14.[001 to 014, 016, 017, 020, 026, 027, 030, 031, 903, 904, 993, 996, 998, 999]-SubfieldCount	Count(Subfields in 14.[001 to 014, 016, 017, 020, 026, 027, 030, 031, 903, 904, 993, 996, 998, 999]) EQ 1					T
			1	M	14.[001 to 012, 016, 017, 020, 026, 027, 030, 031, 903, 993, 996, 999]-InfoItemCount	Count(InfoItems in Subfield:1 in 14.[001 to 012, 016, 017, 020, 026, 027, 030, 031, 903, 993, 996, 999]) EQ 1					T
			1	M	14.013-InfoItemCount	ForEach(Subfield in 14.013) { Count(InfoItems in Subfield) EQ 1 }					T

ID	Type	Mult	Definition		T
					T
14.014-InfoItemCount	D	1	Count(InfoItems in 14.014) EQ 2		T
14.015-SubfieldCount	D	1	<See Requirement ID: "Field: PPC-Subfield Occurrence">	t-2	T
14.018-SubfieldCount	O	1	Count(Subfields in 14.018) MO [1 to 5]		T
14.018-InfoItemCount	O	1	ForEach(Subfield in 14.018) { Count(InfoItems in Subfield) EQ 2 }		T
14.021-SubfieldCount	D	1	Count(Subfields in 14.021) MO [1 to 5]		T
14.021-InfoItemCount	D	1	ForEach(Subfield in 14.021) { Count(InfoItems in Subfield) EQ 5 }		T
14.022-SubfieldCount	O	1	Count(Subfields in 14.022) MO [1 to 5]		T
14.022-InfoItemCount	O	1	ForEach(Subfield in 14.022) { Count(InfoItems in Subfield) EQ 2 }		T
14.023-SubfieldCount	O	1	Count(Subfields in 14.023) MO [1 to 5]		T
14.023-InfoItemCount	O	1	ForEach(Subfield in 14.023) { Count(InfoItems in Subfield) EQ 4 }		T
14.024-SubfieldCount	O	1	Count(Subfields in 14.024) MO [1 to 5]		T
14.024-InfoItemCount	O	1	ForEach(Subfield in 14.024) { Count(InfoItems in Subfield) EQ 4 }		T
14.025-SubfieldCount	O	1	Count(Subfields in 14.025) MO [1 to 5]		T
14.025-InfoItemCount	O	1	ForEach(Subfield in 14.025) {		T

				unt	Count(InfoItems in Subfield) EQ 2 + 2*{InfoItem:2 in Subfield} }		T
	1	O	14.902-SubfieldCount	Count(Subfields in 14.902) GTE 1			T
	1	O	14.902-InfoItemCount	ForEach(Subfield in 14.902) { Count(InfoItems in Subfield) EQ 4 }			T
	1	O	14.904-SubfieldCount	Count(InfoItems in 14.904) EQ 3			T
	1	O	14.995-SubfieldCount	Count(Subfields in 14.995) MO [1 to 255]			T
	1	O	14.995-InfoItemCount	ForEach(Subfield in 14.995) { Count(InfoItems in Subfield) EQ 1 OR 2 }			T
	1	O	14.997-SubfieldCount	Count(Subfields in 14.997) MO [1 to 255]			T
	1	O	14.997-InfoItemCount	ForEach(Subfield in 14.997) { Count(InfoItems in Subfield) EQ 1 OR 2 }			T
	1	O	14.998-SubfieldCount	<See Requirement ID: _"Field: Geographic"_>	t-2		
Field: Type14-CondCode	Table 71 <Table 71 specifies the Condition Code for each field.>	1	-	[14.001 to 14.005, 14.013]-Mandatory CondCode	Present(14.001 to 14.005, 14.013)		B
		1	-	[14.019, 14.028, 14.029, 14.032 to 14.199, 14.901, 14.905 to 14.992, 14.994]-Reserved	NOT Present(14.019, 14.028, 14.029, 14.032 to 14.199, 14 901, 14.905 to 14 992, 14.994)		B
Record:	Table 71, This field is mandatory if an image is	2	D	[14.006 to	Present(14.999) IFF		B

Field	Reference			Content / Condition		
14.006 to 14.012 Dependent	8.14.6 to 8.14.12		14.012]-CondCode Dependent	present in Field 14.999. Otherwise it is absent. / Present(14.006 to 14.012)		
Field: 14.014-Print Position Descriptors Dependent	Table 71, 8.14.14	D	14.014-CondCode Dependent	This field shall be present if and only if the finger position code "19" appears in Field 14.013: Friction ridge generalized position / FGP. / See Requirement ID: "Field: PPD Conditional"	t-2	
Field: 14.015-Print Position Coordinates Dependent	Table 71, 8.14.15	D	14.015-CondCode Dependent	This field may be present if and only if the finger position code "19" appears in Field 14.013: Friction ridge generalized position / FGP. / See Requirement ID: "Field: PPD Conditional"	t-2	
Field: 14.021-Finger Segment Position Dependent	Table 71, 8.14.20	2 / D	14.021-CondCode Dependent	This optional field shall contain offsets to the locations of image segments containing the individual fingers within the flat images of simultaneous fingers from each hand or the two simultaneous thumbs. (FGP = 13, 14, 15 or 40-50 from Table 8 as entered in Field 14.013: Friction ridge generalized position / FGP). / IF Present(14.021) THEN {InfoItem:1 in Subfield:1 in 14.013} MO [13 to 15, 40 to 50]		B*
Field: 14.027-Stitched Image Flag Dependent	Table 71, 8.14.26	3 / D	14.027-CondCode Dependent	This field signifies that images captured separately were stitched together to form a single image. This field is mandatory if an image has been stitched, and the value shall be set to 'Y'. Otherwise, this field shall not appear in the record. / <Unsupported: It is not possible to easily detect if the image was stitched.>	t-1	
Field: 14.999-Palmprint Image Dependent	Table 71, 8.14.38	2 / D	14.999-CondCode Dependent	This field contains the palmprint image. It shall contain an image, unless Field 14.018: Amputated or bandaged / AMP has a value or "UP". In the latter case, the field is optional. / IF {InfoItem:2 in All{Subfield in 14.018}} NEQ ASCII(UP) THEN Present(14.999)		
Field: Type14-CharType	8.14, Table 71	1 / -	14.[001,002,003,006 to 010, 012,016, 017, 026, 031]-CharType	<Table 71 specifies the Character Type for each field that contains no subfields.> / Bytes(14.[001,002,003, 006 to 010, 012,016, 017, 026, 031]) MO [0x30 to 0x39]		B
		1 / -	14.[027, 030)]-CharType	Bytes(14.[027, 030)]) [0x20, 0x41 to 0x5A, 0x61 to 0x7A]		B
		1 / -	[14.004, 14.020, 14.993]-CharType	TRUE		B
		1 / -	14.005-	Bytes(14.005) MO [0x30 to 0x39]		T

126

Field	Ref		#		Type	Description	Code
Field: Type 14-Subfield CharType	8.14, Table 71	<Table 71 specifies the Character Type for each subfield.>			CharType		
			1	-	NIEM-14.005-CharType	Bytes(14.005) MO [0x30 to 0x39, 0x2D]	X
			1	O	14.011-CharType	Bytes(14.011) MO [0x30 to 0x39, 0x20, 0x41 to 0x5A, 0x61 to 0x7A]	B
			1	O	14.020-CharType	TRUE	B
			1	O	14.903-CharType	Bytes(14.903) MO [0x20 to 0x7E]	B
			1	O	14.996-CharType	Bytes(14 996) MO [0x30 to 0x39,0x41 to 0x46, 0x61 to 0x66]	B
			1	O	14.999-CharType	TRUE	B
			1	M	14.013-FGP-CharType	Bytes(All (InfoItems in 14.013)) MO [0x30 to 0x39]	B*
			1	D	14.014-DFP-CharType	ForEach(Subfield in 14.014) { Bytes(InfoItem:1 in Subfield) MO [0x30 to 0x39] }	B*
			1	D	14.014-FIC-CharType	ForEach(Subfield in 14.014) { Bytes(InfoItem:2 in Subfield) MO [0x30 to 0x39, 0x20, 0x41 to 0x5A, 0x61 to 0x7A] }	B*
			1	D	14.015-FVC-CharType	ForEach(Subfield in 14.015) { Bytes((InfoItem:1 in Subfield)) MO [0x30 to 0x39, 0x20, 0x41 to 0x5A, 0x61 to 0x7A] }	B*
			1	D	14.015-LOS-CharType	ForEach(Subfield in 14.015) { Bytes((InfoItem:2 in Subfield)) MO [0x20, 0x41 to 0x5A, 0x61 to 0x7A] }	B*
			1	D	14.015-[LHC, RHC, TVC, BVC]-CharType	ForEach(Subfield in 14.015) { Bytes(InfoItem:3 to 6 in Subfield)) MO [0x30 to 0x39] }	B*
			1	O	14.018-FRAP-CharType	ForEach(Subfield in 14.018) { Bytes(InfoItem:1 in Subfield)) MO [0x30 to 0x39] }	B*

1	O	14.018-ABC-CharType	ForEach(Subfield in 14.018) { Bytes(InfoItem:2 in Subfield)) MO [0x20, 0x41 to 0x5A, 0x61 to 0x7A] }	B*
1	D	14.021-[FRSP, LHC, RHC, TVC, BVC]-CharType	Bytes(All(InfoItems in 14.021)) MO [0x30 to 0x39]	B*
1	O	14.022-[FRNP, IQS]-CharType	Bytes(All(InfoItems in 14.022)) MO [0x30 to 0x39]	B*
1	O	14.023-[FRQP, QVU, QAP]-CharType	ForEach(Subfield in 14.023) { Bytes(InfoItem:1,2,4 in Subfield)) MO [0x30 to 0x39]	B*
1	O	14.023-QAV-CharType	ForEach(Subfield in 14.023) { Bytes(InfoItem:3 in Subfield)) MO [0x30 to 0x39, 0x41 to 0x46, 0x61 to 0x66] }	B*
1	O	14.024-[FRMP, QVU, QAP]-CharType	ForEach(Subfield in 14.024) { Bytes(InfoItem:1,2,4 in Subfield)) MO [0x30 to 0x39] }	B*
1	O	14.024-QAV-CharType	ForEach(Subfield in 14.024) { Bytes(InfoItem:3 in Subfield)) MO [0x30 to 0x39, 0x41 to 0x46, 0x61 to 0x66] }	B*
1	O	14.025-[FRAS, NOP, HPO, VPO]-CharType	Bytes(All(InfoItems in 14.025)) MO [0x30 to 0x39]	B*
1	O	14.902-[NAV, OWN, PRO]-CharType	TRUE	T
1	O	14.902-GMT-	ForEach(Subfield in 14.902) {	T

Field				CharType					
	1	O	NIEM-14.902-SubfieldCharType	Bytes(InfoItem:1 in Subfield) MO [0x30 to 0x39,0x5A] }					X*
	1	O	14.904-[MAK, MOD, SER]-CharType	< The treatment of subfields for validation in the XML version requires further review. Byte values allowed for first "subfield" in XML are 0x30 to 0x39, 0x3A, 0x54, 0x5A.> TRUE					B*
	1	O	14.995-[ACN, ASP]-CharType	Bytes(All(InfoItem:1,2 in 14 995)) MO [0x30 to 0x39]					B*
	1	O	14.997-[SRN, RSP]-CharType	Bytes(All(InfoItem:1,2 in 14 997)) MO [0x30 to 0x39]					B*
	1	O	14.998-[UTE, LTD,LTM, LTS, LGD, LGM, LGS, ELE, GDC, GCM, GCE, GCN, GRT, OSI, OCV]-CharType	<See Requirement ID: "Field: Geographic">	t-2				
Field: Type14-CharCount	Table 71			<Table 71 specifies the Character Ccunt for each field that contains no subfields.>					
	1	M	14.001-CharCount	DataLength(14.001) MO [1 to 8]					T
	1	M	NIEM-14.001-CharCount	Length(14.001) EQ 2					X
	1	M	14.002-CharCount	DataLength(14.002) EQ 1 OR 2					B
	1	M	14.003-CharCount	DataLength(14.003) EQ 1 OR 2					B
	1	M	14.004-CharCount	<See Requirement ID: "Field: Originating Agency".>	t-2				
	1	M	14.005-CharCount	DataLength(14.005) EQ 8					T
	1	M	Niem-14.005-CharCount	DataLength(14.005) EQ 10					X
	1	M	14.006-CharCount	DataLength(14.006) MO [2 to 5]					B

1	M	14.007-CharCount	DataLength(14.007) MO [2 to 5]	B
1	M	14.008-CharCount	DataLength(14.008) EQ 1	B
1	M	14.009-CharCount	DataLength(14.009) MO [1 to 5]	B
1	M	14.010-CharCount	DataLength(14.010) MO [1 to 5]	B
1	M	14.011-CharCount	DataLength(14.011) MO [3 to 5]	B
1	M	14.012-CharCount	DataLength(14.012) EQ 1 OR 2	B
1	O	14.016-CharCount	DataLength(14.016) MO [1 to 5]	B
1	O	14.017-CharCount	DataLength(14.017) MO [1 to 5]	B
1	O	14.020-CharCount	DataLength(14.020) MO [1 to 126]	B
1	O	14.026-CharCount	DataLength(14.026) MO [1 to 3]	B
1	D	14.027-CharCount	DataLength(14.027) EQ 1	B
1	O	14.030-CharCount	DataLength(14.030) MO [7 to 10]	B
1	O	14.031-CharCount	DataLength(14.031) EQ 2	B
1	O	14.903-CharCount	DataLength(14.903) MO [13 to 16]	B
1	O	14.993-CharCount	<See Requirment ID: "Field: Source Agency Name".> t-2	
1	O	14.996-CharCount	DataLength(14.995) EQ 64	B
1	M	14.999-CharCount	DataLength(14.999) GTE 1	B
1	M	14.013-FGP-CharCount	Length(All(InfoItems in 14.013)) EQ 1 OR 2	B*
1	D	14.014-DFP-CharCount	Length(InfoItem:1 in 14.014) EQ 1 OR 2	B*
1	D	14.014-FIC-CharCount	Length(InfoItem:2 in 14.014) EQ 3	B*

Table 71 <Table 71 specifies the Character Count for each subfield.>

Field: Type14-Subfield CharCount

1	D	14.015-FVC-CharCount	ForEach(Subfield in 14.015) { Length(InfoItem:1 in Subfield) EQ 2 OR 3 }	B*
1	D	14.015-LOS-CharCount	ForEach(Subfield in 14.015) { Length(InfoItem:2 in Subfield) EQ 2 OR 3 }	B*
1	D	14.015-[LHC, RHC, TVC, BVC]-CharCount	ForEach(Subfield in 14.015) { Length(InfoItem:3 to 6 in Subfield) MO [1 to 5] }	B*
1	O	14.018-FRAP-CharCount	ForEach(Subfield in 14.018) { Length(InfoItem:1 in Subfield) EQ 1 OR 2 }	B*
1	O	14.018-ABC-CharCount	ForEach(Subfield in 14.018) { Length(InfoItem:2 in Subfield) EQ 2 }	B*
1	D	14.021-FRSP-CharCount	ForEach(Subfield in 14.021) { Length(InfoItem:1 in Subfield) EQ 1 OR 2 }	B*
1	D	14.021-[LHC, RHC, TVC, BVC]-CharCount	ForEach(Subfield in 14.021) { Length(InfoItem:2 to 5 in Subfield) MO [1 to 5] }	B*
1	O	14.022-FRNP-CharCount	ForEach(Subfield in 14.022) { Length(InfoItem:1 in Subfield) EQ 1 OR 2 }	B*
1	O	14.022-IQS-CharCount	ForEach(Subfield in 14.022) { Length(InfoItem:2 to 5 in Subfield)) MO [1 to 3] }	B*
1	O	14.023-	ForEach(Subfield in 14.023)	B*

Field	M/O	Count	Condition	
FRQP-CharCount			{ Length(InfoItem:1 in Subfield) EQ 1 OR 2 }	B*
14.023-QVU-CharCount	O	1	ForEach(Subfield in 14.023) { Length(InfoItem:2 in Subfield) MO [1 to 3] }	B*
14.023-QAV-CharCount	O	1	ForEach(Subfield in 14.023) { Length(InfoItem:3 in Subfield) EQ 4 }	B*
14.023-QAP-CharCount	O	1	ForEach(Subfield in 14.023) { Length(InfoItem:4 in Subfield) MO [1 to 5] }	B*
14.024-FRMP-CharCount	O	1	ForEach(Subfield in 14.024) { Length(InfoItem:1 in Subfield) EQ 1 OR 2 }	B*
14.024-QVU-CharCount	O	1	ForEach(Subfield in 14.024) { Length(InfoItem:2 in Subfield) MO [1 to 3] }	B*
14.024-QAV-CharCount	O	1	ForEach(Subfield in 14.024) { Length(InfoItem:3 in Subfield) EQ 4 }	B*
14.024-QAP-CharCount	O	1	ForEach(Subfield in 14.024) { Length(InfoItem:4 in Subfield) MO [1 to 5] }	B*
14.025-[FRAS, NOP]-CharCount	O	1	ForEach(Subfield in 14.025) { Length(InfoItem:1,2 in Subfield) EQ 1 OR 2 A}	B*
14.025-[HPO, VPO]-	O	1	ForEach(Subfield in 14.025) ForEach(InfoItem in Subfield ST InfoItem NOT	B*

	CharCount	InfoItem:1 OR InfoItem 2 in Subfield) Length(InfoItem) MO [1 to 5]		
1	O	14.902-GMT-CharCount	ForEach(Subfield in 14.902) { Length(InfoItem:1 in Subfield) EQ 15 }	T
1	O	14.902-[NAV, OWN]-CharCount	ForEach(Subfield in 14.902) { Length(InfoItem:2,3 in Subfield) MO [1 to 64] }	T
1	O	14.902-[PRO]-CharCount	ForEach(Subfield in 14.902) { Length(InfoItem:4 in Subfield) MO [1 to 255] }	T
1	O	NIEM-14.902-Subfield-CharCount	< The treatment of subfields for validation in the XML version requires further review. Length of the first "subfield" in XML is 20.>	X*
1	O	14.904-[MAK, MOD, SER]-CharCount	Length(All(InfoItems in 14.904) MO [1 to 50]	B*
1	O	14.995-ACN-CharCount	ForEach(Subfield in 14.995) { Length(InfoItem:1 in Subfield) MO [1 to 3] }	B*
1	O	14.995-ASP-CharCount	ForEach(Subfield in 14.995) { Length(InfoItem:2 in Subfield) EQ 1 OR 2 }	B*
1	O	14.997-SRN-CharCount	ForEach(Subfield in 14.997) { Length(InfoItem:1 in Subfield) MO [1 to 3] }	B*
1	O	14.997-RSP-CharCount	ForEach(Subfield in 14.997) { Length(InfoItem:2 in Subfield) EQ 1 OR 2 }	B*
	O	14.998-[UTE, LTD,LTM, LTS, LGD, LGM, LGS,	<See Requirement ID: "Field: Geographic" >	t-2

133

Field	Reference	Description	Occ		Field Identifier	Requirement / Condition		Code
Field: Type14-Field Occurrence	Table 71	<Table 71 specifies the Field Occurrence for each field.>	1	-	14.[019,028,029,032 to 199,901,905 to 992,994]-Occurrence	Count(14.[019,028,029,032 to 199,901,905 to 992,994]) EQ 0		B
			1	M	14.[001 to 005,013]-Occurrence	Count(14.[001 to 005,013]) EQ 1		B
			1	-	14.[006 to 012,014 to 018,020 to 027,030,031,902 to 904,993,995 to 999]-Occurrence	Count(14.[006 to 012,014 to 018,020 to 027,030,031,902 to 904,993,995 to 999]) LTE 1		B
Field: 14.001-Record Header Value	8.14.1, Table 71, 7.1	Field 14.001 Record header. In Traditional encoding, this field contains the record length in bytes (including all information separators)	M		14.001-Record Header	<See Requirement ID "Field: xx.001-Record Header">	t-2	
Field: 14.001-Value	8.14.1, C.9.12	The XML name for the Type-14 record is <itl:PackageFingerprintImageRecord>, and its <biom:RecordCategoryCode> elements shall have a value of "14".	1	M	NIEM-14.001-Value	ForEach(XElm(itl:PackageFingerprintImageRecord){ {XElm(biom:RecordCategoryCode)} EQ ASCII(14) }		X
Field: 14.002-Information Designation Character Value	8.14.2, Table 71, 7.3.1	This mandatory field shall be the IDC of this Type-14 record as found in the information item IDC of Field 1.003 Transaction content/CNT.	1	M	14.002-IDC	<See Requirement IDs "Field: xx.002-IDC "and "Field: 1.003-Transaction Content Subfield 2 IDC Matches" >	t-2	
Field: 14.003-Impression Type Value	8.14.3, Table 71, 7.7.4.1	This mandatory field shall indicate the manner by which the latent print was obtained. See Section 7.7.4.1 for details. <Table 71 lists the valid values for IMP.>	1	M	14.003-Value	{14.003} MO [0 to 3, 8, 20 to 29] MO [Integers]		B
Field:	8.14.4,	This is a mandatory field. See Section 7.6	M		14.004-	<See Requirement ID: "Field: Originating	t-2	

Field	Reference	Description	#	Mnd	Value	Requirement	Code	Type
14.004-Originating Agency Value	7.6	for details.			Value	Agency">.		
Field: 14.005-Fingerprint Capture Date Value	8.14.5, 7.7.2.3	This mandatory field shall contain the date that the fingerprint data contained in the record was captured.	1	M	14.005-Value	{14.005} MO [ValidLocalDate]	t-6	T
			1	M	NIEM-14.005-Value	ForEach(XElm(iti:PackageFingerprintRecord)) { XElm(nc:Date) in XElm(biom:CaptureDate)} MO [NIEM-ValidLocalDate] }	t-6	X
Field: 14.006-Horizontal Line Length Value	8.14.6, Table 71, 7.7.8.1	The maximum horizontal size is limited to 65,534 pixels in Record Types-4 and 8, and to 99,999 for other record types. The minimum value is 10 pixels.		M	14.006-Value	<See Requirement ID "Field: Image HLL Value" >	t-2	
Field: 14.006-Horizontal Line Length Metadata	8.14.6, Table 71, 7.7.8.1	<The HLL is verified by checking the image metadata if compression is used.>	2	M	14.006-Matches Image Metadata	IF {14.011} EQ ASCII(JPEGB) OR ASCII(JPEGL) THEN {14.006} EQ {ImageWidth-JPEGB,JPEGL} ELSE IF {14.011} EQ ASCII(JP2) OR ASCII(JP2L) THEN {14.006} EQ {ImageWidth-JP2,JP2L} ELSE IF {14.011} EQ ASCII(PNG) THEN {14.006} EQ {ImageWidth-PNG} ELSE IF {14.011} EQ ASCII(WSQ20) THEN {14.006} EQ {ImageWidth-WSQ}	t-11	B
Field: 14.007-Vertical Line Length Value	8.14.7, Table 71, 7.7.8.2	The maximum vertical size is limited to 65,534 pixels in Record Types-4 and 8, and to 99,999 for other record types. The minimum value is 10 pixels.		M	14.007-Value	<See Requirement ID "Field: Image VLL Value" >	t-2	
Field: 14.007-Vertical Line Length Metadata	8.14.7, Table 71, 7.7.8.2	<The VLL is verified by checking the image metadata if compression is used.>	2	M	14.007-Matches Image Metadata	IF {14.011} EQ ASCII(JPEGB) OR ASCII(JPEGL) THEN {14.007} EQ {ImageHeight-JPEGB,JPEGL} ELSE IF {14.011} EQ ASCII(JP2) OR ASCII(JP2L) THEN {14.007} EQ {ImageHeight-JP2,JP2L} ELSE IF {14.011} EQ ASCII(PNG) THEN {14.007} EQ {ImageHeight-PNG} ELSE IF {14.011} EQ ASCII(WSQ20) THEN {14.007} EQ {ImageHeight-WSQ}	t-11	B
Field: 14.008-Scale	8.14.8, Table 71,	<Table 71 lists the value constraints for SLC>		M	14.008-Value	<See Requirement ID "Field: Image SLC Value" >	t-2	

Field	Reference	Description		Char	Requirement Name	Requirement Logic	Test	Type
Units Value								
Field: 14.008- Scale Units Metadata	7.7.8.3 8.14.8, Table 71, 7.7.8.3	A value of "1" shall indicate pixels per inch. A value of "2" shall indicate pixels per centimeter. A value of "0" in this field indicates that no scale is provided, and the quotient of THPS/TVPS shall provide the pixel aspect ratio. <The SLC is verified by checking the image metadata if compression is used.>	2	M	14.008- Matches Image Metadata	IF {14.011} EQ ASCII(JPEGB) OR ASCII(JPEGL) THEN {14.008} EQ {SamplingUnits-JPEGB,JPEGL} ELSE IF {14.011} EQ ASCII(JP2) OR ASCII (JP2L) OR ASCII (WSQ20) THEN <Provide Warning "Not Tested"> ELSE IF {14.011} EQ ASCII(PNG) THEN IF {14.008} EQ 1 OR 2 THEN { SamplingUnits-PNG} EQ 1, ELSE IF {14.008} EQ 0 THEN { SamplingUnits-PNG} EQ 0	t-11	B
Field: 14.009- Transmitted Horizontal Pixel Scale Value	8.14.9, Table 71, 7.7.8.4	<Table 71 lists the value constraints for THPS.>		M	14.009- Value	<See Requirement ID "Field: Image THPS Value" >	t-2	
Field: 14.009- Transmitted Horizontal Pixel Scale Metadata	8.14.9, Table 71, 7.7.8.4	This is the integer pixel density used in the horizontal direction of the image if SLC has a value of "1" or "2". If SLC has a value of "0", this information item shall contain the horizontal component of the pixel aspect ratio, up to 5 digits. <The THPS is verified by checking the image metadata if compression is used >	2	M	14.009- Matches Image Metadata	IF {14.011} EQ ASCII(JPEGB) OR ASCII(JPEGL) AND {14.008} EQ 1 OR 2 THEN {14.009} EQ {HorizontalDensity-JPEGB,JPEGL} ELSE IF {14.011} EQ ASCII(JP2) OR ASCII (JP2L) OR ASCII (WSQ20) THEN <Provide Warning "Not Tested"> ELSE IF {14.011} EQ ASCII(PNG) AND {14.008} EQ 1 THEN {14.009} EQ {HorizontalDensity-PNG} * 0.0254 (meters/inch), ELSE IF {14.011} EQ ASCII(PNG) AND {14.008} EQ 2 THEN {14.009} EQ {HorizontalDensity-PNG} * 0.01 (meters/cm)	t-11, t-12	B
			2	M	14.009- Aspect Ratio Matches Image Metadata	IF {14.011} EQ ASCII(JPEGB) OR ASCII(JPEGL) AND {14.008} NEQ 1 OR 2 THEN {14.009}/{14.010} EQ {HorizontalDensity-JPEGB,JPEGL} / {VerticalDensity-JPEGB,JPEGL} ELSE IF {14.011} EQ ASCII(JP2) OR ASCII (JP2L) OR ASCII (WSQ20) THEN <Provide Warning "Not Tested"> ELSE IF {14.011} EQ ASCII(PNG) AND {14.008} NEQ 1 OR 2 THEN {14.009}/{14.010} EQ {HorizontalDensity-PNG} / {VerticalDensity-PNG}	t-11	B
Field: 14.010-	8.14.10,	<Table 71 lists the value constraints for		M	14.010-	<See Requirement ID "Field: Image TVPS Value"	t-2	

136

Field	Reference	Description		Value	Condition	Code	
14.010-Transmitted Vertical Pixel Scale Value	Table 71, 7.7.8.5	TVPS.>		Value	>		
Field: 14.010-Transmitted Vertical Pixel Scale Metadata	8.14.10, Table 71, 7.7.8.5	This is the integer pixel density used in the Vertical direction of the image if SLC has a value of "1" or "2". If SLC has a value of "0", this information item shall contain the Vertical component of the pixel aspect ratio, up to 5 digits. <The TVPS is verified by checking the image metadata if compression is used >	2 / M	14.010-Matches Image Metadata	IF {14.011} EQ ASCII(JPEGB) OR ASCII(JPEGL) AND {14.008} EQ 1 OR2 THEN {14.010} EQ {VerticalDensity-JPEGB,JPEGL} ELSE IF {14.011} EQ ASCII(JP2) OR ASCII (JP2L) OR ASCII (WSQ20) THEN <Provide Warning "Not Tested"> ELSE IF {14.011} EQ ASCII(PNG) AND {14.008} EQ 1 THEN {14.010} EQ {VerticalDensity-PNG} * 0.0254 (meters/inch), ELSE IF {14.011} EQ ASCII(PNG) AND {14.008} EQ 2 THEN {14.010} EQ {VerticalDensity-PNG} * 0.01 (meters/cm)	t-11, t-12	B
			2 / M	14.010-Aspect Ratio Matches Image Metadata	IF {14.011} EQ ASCII(JPEGB) OR ASCII(JPEGL) AND {14.008} NEQ 1 OR 2 THEN {14.009}/{14.010} EQ {HorizontalDensity-JPEGB,JPEGL} / {VerticalDensity-JPEGB,JPEGL} ELSE IF {14.011} EQ ASCII(JP2) OR ASCII (JP2L) OR ASCII (WSQ20) THEN <Provide Warning "Not Tested"> ELSE IF {14.011} EQ ASCII(PNG) AND {14.008} NEQ 1 OR 2 THEN {14.009}/{14.010} EQ {HorizontalDensity-PNG} / {VerticalDensity-PNG}	t-11	B

Field: 14.011-Compression Algorithm Value	8.14.11, Table 71, 7.7.9.1	For each of these fields, the entry corresponds to the appropriate *Label* entry in Table 15: Field 14.011: Compression algorithm / CGA.
		
		Wavelet Scalar Quantization (WSQ) shall be used for compressing grayscale friction ridge data at 500 ppi class for new systems. In order to maintain backward compatibility, legacy systems may use JPEGB or JPEGL for compressing 500 ppi class images.
		
		WSQ shall not be used for other than the
			2 / M	14.011-Value	<Note: 19.49 and 19.89 are ppmm. The units for 14.016 and 14.017 depend on 14.008.>
					
					IF {14.016} AND {14.017} GTE 19.49 AND LTE 19.89 THEN {
					IF {14.011} MO [ASCII(JPEGB, JPEGL)] THEN <Provide Legacy Warning> ELSE {14.011} EQ ASCII(WSQ20) }
					
					ELSE IF {14.016} AND {14.017} GTE 38.98 AND LTE 39.76 {
							B

Field	Reference	Description	#	M/D	Value	Logic	Code	

		500 ppi class.
		
		For friction ridge images at the 1000 ppi class, JPEG 2000 shall be used according to the specifications and options contained in Profile for 1000 ppi Fingerprint Compression.
						{14.011} EQ [ASCII(JP2,JP2L)] }
						
						ELSE
						{
						{14.011} MO [ASCII(NONE, JPEGB, JPEGL, JP2, JPEGL JP2, JP2L, PNG)] }
								B

Field: 14.011-Compression Algorithm Metadata	8.14.11, Table 71	<The CGA is verified by checking the image metadata for the compression type signature if compression is used.>	2	M	14.011-Matches Image Metadata	IF {14.011} EQ ASCII(JPEGB) OR ASCII(JPEGL) THEN Present(SOI-JPEG,JPEGL) ELSE IF {14.011} EQ ASCII(JP2) OR ASCII(JP2L) THEN Present(SigBox) ELSE IF {14.011} EQ ASCII(PNG) THEN Present(PNGSig) ELSE IF {14.011} EQ ASCII(WSQ20) THEN Present(SOI-WSQ)	t-11	
Field: 14.012-Bits Per Pixel Value	8.14.12, Table 71, 7.7.8.6	This field shall contain an entry of "8" for normal grayscale values of "0" to "255". Any entry in this field greater than "8" shall represent a grayscale pixel with increased proportion.		M	14.012-Value	<See Requirement ID "Field: Image BPX Value" >	t-2	
Field: 14.012-Bits Per Pixel Metadata	8.14.12, Table 71	<The BPX is verified by checking the image metadata for the compression type signature if compression is used.>	2	M	14.012-Matches Image Metadata	IF {14.011} EQ ASCII(JPEGB) OR ASCII(JPEGL) THEN {14.012} EQ {BPX-JPEG, JPEGL} ELSE IF {14.011} EQ ASCII(JP2) OR ASCII(JP2L) THEN {14.012} EQ {BPX-JP2,JP2L} ELSE IF {14.011} EQ ASCII(PNG) THEN {14.012} EQ {BPX-PNG} ELSE <Provide Warning "Not Tested">	t-11	B
Field: 14.013-Friction Ridge Generalized Position Value	8.14.13, Table 71, 7.7.4.2, Table 8	See Section 7.7.4.2 for details.	1	M	14.013-FGP-Value	[InfoItem: 1 in Subfield: 1 in 14.013] MO [0 to 19, 33, 36, 40 to 50] AND MO [Integers]		B
Field: 14.014-Print Position Descriptors	8.14.14, Table 71, 7.7.4.3	...described in Section 7.7.4.3		D	14.014-[DFP, FIC]-Value	<See Requirement ID: "Field: SPD PPD Values">	t-2	

Field	Reference	Description	#	M/D/O	Field ID	Requirement		B*
Value								
Field: 14.014- Print Position Descriptors Conditional	8.14.14, Table 71	This field shall be present if and only if the finger position code "19" appears in Field 14.013: Friction ridge generalized position / FGP.		D	14.014-Conditional	<See Requirement ID: "Field: PPD Conditional">	t-2	
Field: 14.015- Print Position Coordinates Value	8.14.15, Table 71, 7.7.4.4	See section 7.7.4.4		D	14.015-[FVC, LOS, LHC, RHC, TVC, BVC]-Value	<See Requirement IDs: "Field: PPC-Subfield 1" through "Field: PPC-SubfieldCount 5-6".>	t-2	
Field: 14.015- Print Position Coordinates Conditional	8.14.15, Table 71	This field may be present if and only if the finger position code "19" appears in Field 14.013: Friction ridge generalized position / FGP.		D	14.015-Conditional	<See Requirement ID: "Field: SPD PPC Conditional">	t-2	
Field: 14.016- Scanned Horizontal Pixel Scale Value	8.14.16, Table 71, 7.7.8.7	See section 7.7.8.7 for details.		O	14.016-Value	<See Requirement IDs: "Field: Image SHPS Value">	t-2	
Field: 14.017- Scanned Vertical Pixel Scale Value	8.14.17, Table 71, 7.7.8.8	See section 7.7.8.8 for details.		O	14.017-Value	<See Requirement IDs: "Field: Image SVPS Value">	t-2	
Field: 14.018- Amputated Or Bandaged Value	8.14.18, Table 71, Table 72	<Table 71 lists the value constraints for AMP.>	1	M ⇑	14.018-FRAP-Value	ForEach(Subfield in 14.018) { {InfoItem:1 in Subfield} MO [1 to 10, 16,17] }		B*
			1	M ⇑	14.018-ABC-Value	ForEach(Subfield in 14.018) { {InfoItem:2 in Subfield} MO [ASCII(XX, UP)] }		B*
Field: 14.019- Reserved	Table 71	Reserved for future use only by ANSI/NIST-ITL.		-	14.019-Reserved	<See Requirement ID: "Field: Type14 CondCode>.	t-2	
Field: 14.020- Comment Value	8.14.19, Table 71, 7.4.4	See section 7.4.4 for details.		O	14.020-Value	<See Requirement ID: "Field: Comment>.	t-2	
Field: 14.021- Fingerprint	8.14.20, Table 71, Table 8	<Table 71 lists the value constraints for SEG.>	1	M ⇑	14.021-FRSP-Value	ForEach(Subfield in 14.021) { {InfoItem:1 in Subfield} MO [1 to 10, 16,17] }		B*

139

Field	Ref	Description	Occ	M/D/O	Subfield	Condition	Type
Segment Position Value			2	M ⇑	14.021-LHC-Value	} ForEach(Subfield in 14.021) { [InfoItem:2 in Subfield] GTE 0 AND LTE {14.006} AND MO [Integers] }	B*
			2	M ⇑	14.021-RHC-Value	ForEach(Subfield in 14.021) { [InfoItem:3 in Subfield] LTE {14.006} AND GT [InfoItem:2 in 14.021} MO [Integers] }	B*
			2	M ⇑	14.021-TVC-Value	ForEach(Subfield in 14.021) { [InfoItem:4 in Subfield] GTE 0 AND LTE {14.007} AND MO [Integers] }	B*
			2	M ⇑	14.021-BVC-Value	ForEach(Subfield in 14.021) { [InfoItem:5 in Subfield] LTE {14.007} AND GT [InfoItem:4 in 14.021} MO [Integers] }	B*
Field: 14.021-Fingerprint Segment Position Conditional	8.14.20, Table 71	This optional field shall contain offsets to the locations of image segments containing the individual fingers within the flat images of simultaneous fingers from each hand or the two simultaneous thumbs. (FGP = 13, 14, 15 or 40-50 from Table 8 as entered in Field 14.013: Friction ridge generalized position / FGP).		D	14.021-Conditional	<See Requirement ID: "Field: 14.021-Finger Segment Position Dependent">	t-2
Field: 14.022-NIST Quality Metric Value	8.14.21, Table 71, Table 86	<Table 71 lists the value constraints for NQM.>	1	M ⇑	14.022-FRNP-Value	ForEach(Subfield in 14.022) { [InfoItem:1 in Subfield] MO [1 to 10, 16,17] }	B*
			1	M ⇑	14.022-IQS-Value	ForEach(Subfield in 14.022) { [InfoItem:2 in Subfield] MO [1 to 5, 254, 255] MO [Integers] }	B*
Field: 14.023-Segmentation Quality Metric Value	8.14.22, Table 71, Table 68	<Table 71 lists the value constraints for SQM.>		M ⇑	14.023-[FRQP, QVU, QAV, QAP]-Value	<See Requirement ID: "Field: Sample Quality Occurrences", "Field: Sample Quality Subfield 1", "Field: Sample Quality Subfield 2", and "Field: Sample Quality Subfield 3".>	t-2
Field: 14.024-	8.14.23, Table 71,	<Table 71 lists the value constraints for FQM.>		O	14.024-[FRMP,	<See Requirement ID: "Field: Sample Quality Occurrences", "Field: Sample Quality Subfield 1",	t-2

Field	Reference	Description	Count	M/O/D	QVU, QAV, QAPI-Value	Value Constraint	Note	Status
Finger Quality Metric Value	Table 8					"Field: Sample Quality Subfield 2", and "Field: Sample Quality Subfield 3".>		B*
Field: 14.025-Alternate Finger Segment Position(s) Value	8.14.24, Table 71, Table 8	<Table 71 lists the value constraints for ASEG.>	1	M ⇐	14.025-FRAS-Value	ForEach(Subfield in 14.025) { {InfoItem:1 in Subfield} MO [1 to 10, 16, 17] }		B*
			1	M ⇐	14.025-NOP-Value	ForEach(Subfield in 14.025) { {InfoItem:2 in Subfield} MO [3 to 99] }		B*
			2	M ⇐	14.025-[HPO, VPO]-Value	ForEach(Subfield in 14.025) { For(X EQ 3 to {InfoItem:2 in Subfield}) { IF X MOD 2 EQ 0 {InfoItem:X in Subfield} GTE 0 AND LTE {14.007} MO [Integers] ELSE {InfoItem:X in Subfield} GTE 0 AND LTE {14.006} MO [Integers] } }		B*
Field: 14.025-Alternate Finger Segment Position(s) Polygon	8.14.24, Table 71, Section 7.8	No two vertices may occupy the same location.	2	O	14.025-ASEG Vertices unique	<Each Vertex (X, Y) is unique>		B*
		The order of the vertices shall be in their consecutive order around the perimeter of the polygon, either clockwise or counterclockwise. The polygon side defined by the last vertex and the first vertex shall complete the polygon. The polygon shall be a simple, plane figure with no sides crossing and no interior holes.	2	O	14.025-ASEG Polygon	<Unsupported.>	t-14	
Field: 14.026-Simultaneous Capture Value	8.14.25, Table 71	<Table 71 lists the value constraints for SCF.>	1	O	14.026-Value	{14.026} MO [1 to 255] MO [Integers]		B
Field: 14.027-Stitched Image Flag	8.14.26, Table 71	<Table 71 lists the value constraints for SCF.>	1	D	14.027-Value	{14.027} EQ ASCII(Y)		B

Value								
Field: 14.027- Stitched Image Flag Conditional	8.14.26, Table 71	This field signifies that images captured separately were stitched together to form a single image. This field is mandatory if an image has been stitched, and the value shall be set to 'Y'. Otherwise, this field shall not appear in the record.	D	14.027- Conditional	<See Requirement ID: "Field: 14.027-Stitched Image Flag Dependent">	t-2		
Field: 14.028, 14.029- Reserved	Table 71	Reserved for future use only by ANSI/NIST-ITL.	-	14.028, 14.029 Reserved	<See Requirement ID: "Field: Type14-CondCode>.	t-2		
Field: 14.030- Device Monitoring Mode Value	8.14.27, Table 71	<Table 71 lists the value constraints for DMM.>	1	14.030- Value	{14.030} MO ASCII(CONTROLLED, ASSISTED, OBSERVED, UNATTENDED, UNKNOWN)			B
Field: 14.031- Subject Acquisition Profile-Fingerprint Value	8.14.28, Table 71	<Table 71 lists the value constraints for FAP.>	1	14.031- Value	{14.031} MO [10, 20, 30, 40, 45, 50, 60] MO [Integers]			B
Field: 14.032 to 14.199- Reserved	Table 71	Reserved for future use only by ANSI/NIST-ITL.	-	14.032 to 14.199 Reserved	<See Requirement ID: "Field: Type14-CondCode>.	t-2		
Field: 14.200 to 14.900- User Defined	Table 71	User Defined Fields	-	14.200 to 14.900- User Defined	TRUE			B
Field: 14.901- Reserved	Table 71	Reserved for future use only by ANSI/NIST-ITL.	-	14.901- Reserved	<See Requirement ID: "Field: Type14-CondCode>.	t-2		
Field: 14.902- Annotated Information Value	8.14.30, Table 71	This is an optional field, listing the operations performed on the original source in order to prepare it for inclusion in a biometric record type. See Section 7.4.1.	O	14.902- [GMT, NAV, OWN, PROJ-Value	<See Requirement ID: "Field: xx.902-ANN" >.	t-2		
Field: 14.903- Device Unique Identifier Value	8.14.31, Table 71	This is an optional field. See Section 7.7.1.1.	O	14.903- Value	<See Requirement ID: "Field: Device ID" >.	t-2		
Field: 14.904-	8.14.32, Table 71	This is an optional field. See Section 7.7.1.2.	O	14.904- [MAK,	<See Requirement ID: "Field: Make Model" >.	t-2		

Make/Model/Serial Number Value

Field	Section	Description			[MOD, SER]-Value	Requirement ID		
Field: 14.905 to 14.992-Reserved	Table 71	Reserved for future use only by ANSI/NIST-ITL.		-	14.905 to 14.992-Reserved	<See Requirement ID: "Field: Type14-CondCode>.	t-2	
Field: 14.993-Source Agency Name	8.14.33, Table 71	This is an optional field. It may contain up to 125 Unicode characters.		O	14.993-Value	<See Requirment ID: "Field: Source Agency Name".>	t-2	
Field: 14.994-Reserved	Table 71	Reserved for future use only by ANSI/NIST-ITL.		-	14.994-Reserved	<See Requirement ID: "Field: Type14-CondCode>.	t-2	
Field: 14.995-Associated Context Value	8.14.34, Table 71	See Section 7.3.3		O	14.995-[ACN, ASP]-Value	<See Requirement IDs: "Field: xx.995-ASC" and "Field: xx.995-ASC-ACN" and "Field: xx.995-ASC-ASP">.	t-2	
Field: 14.996-Hash Value	8.14.35, Table 71	See Section 7.5.2		O	14.996-Value	<See Requirement ID: "Field: HAS">	t-2	
Field: 14.997-Source Representation Value	8.14.36, Table 71	See Section 7.3.2		O	14.997-[SRN, RSP]-Value	<See Requirement IDs: "Field: xx.997-SOR-SRN" and "Field: xx.997-SOR-RSP">.	t-2	
Field: 14.998-Geographic Sample Acquisition Location Value	8.14.37, Table 71	See Section 7.7.3		O	14.998-[UTE, LTD, LTM, LTS, LGD, LGM, LGS, ELE, GDC, GCM, GCE, GCN, GRT, OSI, OCV]-Value	<See Requirement IDs: "Field: Geographic", "Field: Geographic", "Field: Geographic-Subfield 1" through "Field: Geographic-Values-SubField 15" >.	t-2	
Field: 14.999-Image Data Valid	8.14.38, Table 71	This is a mandatory field contains the image. <The image metadata is checked for validity.>	2	M	14.999-Uncompressed Image Length	IF {14.011} EQ ASCII(NONE) THEN Length(14.999) EQ 14.006} * {14.007}		B
			2	M	14.999-Valid Image Format	IF {14.011} EQ ASCII(JPEGB) OR ASCII(JPEGL) THEN Present(JFIF, SOI-JPEGB, JPEGL,	t-11	B

Requirement ID	Reference in Base Standard	Requirement Summary	Level	Status	Assertion ID	Test Assertion	Test Note	Implementation Support	Supported Range	Test Result	Applicability
Field: 14.999- Image WSQ Version 2.0	7.7.9.1	Only version 3.1 or higher shall be used for compressing grayscale fingerprint data at 500 ppi class with a platen area of 2 inches or greater in height. WSQ 2.0 or higher may be used for 500 ppi class data taken from a platen of less than 2 inches in height. WSQ shall not be used for other than the 500 ppi class.	2	M	14.999- Valid WSQ Encoder Version	SOF-JPEGB, JPEGL, EOI-JPEG, JPEGL) ELSE IF {14.011} EQ ASCII(JP2) OR ASCII(JP2L) THEN Present(SigBox, HeadBox, ImgBox, EOI-JP2,JP2L) ELSE IF {14.011} EQ ASCII(PNG) THEN Present(PNGSig, IHDR, IDAT, IEND) ELSE IF {14.999} EQ ASCII(WSQ20) THEN Present(SOI-WSQ,SOF-WSQ,SOB-WSQ,EOI-WSQ) IF {14.011} EQ ASCII(WSQ20) THEN {Encoder Version} EQ 1 OR 2	t-11, t-16				B

Table 6.12 - Assertions for Record Type 15 - Palm Print Image Record

Requirement ID	Reference in Base Standard	Requirement Summary	Level	Status	Assertion ID	Test Assertion	Test Note	Implementation Support	Supported Range	Test Result	Applicability
8.15: Record Type-15: Palm print image record											
Record: Type15- Palm Print Image	8.15	The Type-15 record shall contain and be used to exchange palm print image data together with fixed and user-defined textual information fields pertinent to the digitized image. The image data shall be acquired directly from a subject using a live-scan device, a palmprint card, or other media that	3	M	Type15- Palm Print Image	<Unsupported: It is not feasible to test if the image represents a palm print or how the image was captured.>	t-1				

			contains the subject using a live-scan device, a palmprint card, or other media that contains the subject's palm and/or wrist prints.					
Record: Type15-Palm Print Area	8.15	3	Any method used to acquire the palm print images shall be capable of capturing a set of images for each hand. This set may include the writer be capable of capturing a set of and the entire area of the full palm extending from the wrist bracelet to the tips of the fingers as one or two scanned images. (See Figure 3) If two images are used to represent the full palm, the lower image shall extend from the wrist bracelet to the top of the interdigital area (third finger joint) and shall include the thenar, and hypothenar areas of the palm. The upper image shall extend from the bottom of the interdigital area to the upper tips of the fingers. This provides an adequate amount of overlap between the two images.	M	Type15- Palm Print Area	t-1	\<Unsupported: It is not feasible to test which area of the palm is represented in the image.\>	T
Field: Type15- Subfield Occurence	Table 73	1	\<Table 73 specifies which fields contain subfields as well as the number of occurrences permitted \>	M	15.[001 to 013, 016, 017, 020, 030, 903, 904, 993, 996, 998, 999]- SubfieldCount		Count(Subfields in 15.[001 to 013, 016, 017, 020, 030, 903, 904, 993, 996, 998, 999]) EQ 1	
		1		M	15.[001 to 005, 013, 016, 017, 020, 030, 903, 993, 996, 999]- InfoItemCount		Count(InfoItems in Subfield:1 in 15.[001 to 005, 013, 016, 017, 020, 030, 903, 993, 996, 999]) EQ 1	T
		1		O	15.018- SubfieldCount		Count(Subfields in 15.018) MO [1 to 9]	T
		1		O	15.018- InfoItemCount		ForEach(Subfield in 15.018) { Count(InfoItems in Subfield) EQ 2 }	T
		1		O	15.024-		Count(Subfields in 15.024) MO [1 to 9]	T

Mnemonic	Min	Max	Condition		Type
SubfieldCount	1		ForEach(Subfield in 15.024) { Count(InfoItems in Subfield) EQ 4 }		T
15.024-InfoItemCount	1	O			
15.902-SubfieldCount	1	O	Count(Subfields in 15.902) GTE 1		T
15.902-InfoItemCount	1	O	ForEach(Subfield in 15.902) { Count(InfoItems in Subfield) EQ 4 }		T
15.904-InfoItemCount	1	O	Count(InfoItems in 15.904) EQ 3		T
15.995-SubfieldCount	1	O	Count(Subfields in 15.995) MO [1 to 255]		T
15.995-InfoItemCount	1	O	ForEach(Subfield in 15.995) { Count(InfoItems in Subfield) EQ 1 OR 2 }		T
15.997-SubfieldCount	1	O	Count(Subfields in 15.997) MO [1 to 255]		T
15.997-InfoItemCount	1	O	ForEach(Subfield in 15.997) { Count(InfoItems in Subfield) EQ 1 OR 2 }		T
15.998-Subfields	1	O	<See Requirement ID: "Field: Geographic">	t-2	
[15.001 to 15.005, 15.013]-Mandatory CondCode	1	-	Present(15.001 to 15.005, 15.013)		B
[15.014, 15.015, 15.019, 15.021 to 15.023, 15.025 to 15.029, 15.031 to 15.199, ...]	1	-	NOT Present(15.014, 15.015, 15.019, 15.021 to 15.023, 15.025 to 15.029, 15.031 to 15.199, 15.901, 15.905 to 15.992, 15.994)		B

Table 73   <Table 73 specifies the Condition Code for each field.>

Field: Type15-CondCode

146

Field	Reference	Description			Identifier	Condition / Value	Code
					15.901, 15.905 to 15.992, 15.994]-Reserved		B
Record: 15.006 to 15.012 Dependent	Table 73, 8.15.6 to 8.15.12	This field is mandatory if an image is present in Field 15.999. Otherwise it is absent.	2	D	[15.006 to 15.012]-CondCode Dependent	Present(15.999) IFF Present(15.006 to 15.012)	B
Field: 15.999-Palmprint Image Dependent	Table 73, 8.15.29	This field contains the palmprint image. It shall contain an image, unless Field 15.018: Amputated or bandaged / AMP has a value or "UP". In the latter case, the field is optional.	2	D	15.999-CondCode Dependent	IF {InfoItem:2 in All(Subfield in 15.018)) NEQ ASCII(UP) THEN Present(15.999)	B
Field: Type15-CharType	8.15, Table 73	<Table 73 specifies the Character Type for each field that contains no subfields.>	1	-	15.[001 to 003, 006 to 010, 012, 013, 016, 017]-CharType	Bytes(15.[001 to 003, 006 to 010, 012, 013, 016, 017]) MO [0x30 to 0x39]	B
			1	-	15.030-CharType	Bytes(15.030) [0x20, 0x41 to 0x5A, 0x61 to 0x7A]	B
			1	-	[15.004, 15.020, 15.993]-CharType	TRUE	
			1	M	15.005-CharType	Bytes(15.005) MO [0x30 to 0x39]	T
			1	M	NIEM-15.005-CharType	Bytes(15.005) MO [0x30 to 0x39, 0x2D]	X
			1	M	15.011-CharType	Bytes(15.011) MO [0x30 to 0x39, 0x20, 0x41 to 0x5A, 0x61 to 0x7A]	B
			1	O	15.025-CharType	Bytes(15.025) MO [0x1E, 0x1F, 0x30 to 0x39]	B
			1	O	15.903-CharType	Bytes(15.903) MO [0x20 to 0x7E]	B
			1	O	15.996-CharType	Bytes(10 996) MO [0x30 to 0x39,0x41 to 0x46, 0x61 to 0x66]	B
Field: Type15-Subfield CharType	8.15, Table 73	<Table 73 specifies the Character Type for each subfield.>	1	O	15.018-FRAP-CharType	ForEach(Subfield in 15.018) { Bytes(InfoItem:1 in Subfield) MO [0x30 to 0x39]	B*
			1	O	15.018-ABC-CharType	ForEach(Subfield in 15.018) { Bytes(InfoItem:2 in Subfield)) MO [0x20, 0x41 to	B*

			0x5A, 0x61 to0x7A] }	
1	O	15.024-[FRMP, QVU, QAP]-CharType	ForEach(Subfield in 15.024) { Bytes(InfoItem:1,2,4 in Subfield) MO [0x30 to 0x39] }	B*
1	O	15.024-QAV-CharType	ForEach(Subfield in 15.024) { Bytes(InfoItem:3 in Subfield)) MO [0x30 to 0x39, 0x41 to 0x46, 0x61 to 0x66] }	B*
1	O	15.902-[NAV, OWN, PROJ]-CharType	TRUE	T
1	O	15.902-GMT-CharType	ForEach(Subfield in 15.902) { Bytes(InfoItem:1 in Subfield) MO [0x30 to 0x39,0x5A] }	T
1	O	NIEM-15.902-Subfield CharType	< The treatment of subfields for validation in the XML version requires further review. Byte values allowed for first "subfield" in XML are 0x30 to 0x39, 0x3A, 0x54, 0x5A.>	X*
1	O	15.904-[MAK, MOD, SER]-CharType	TRUE	B*
1	O	15.995-[ACN, ASP]-CharType	Bytes(All(InfoItem:1,2 in 15 995)) MO [0x30 to 0x39]	B*
1	O	15.997-[SRN, RSP]-CharType	Bytes(All(InfoItem:1,2 in 15 997)) MO [0x30 to 0x39]	B*
1	O	15.998-[UTE, LTD,LTM, LTS, LGD, LGM, LGS, ELE, GDC, GCM, GCE, GCN, GRT, OSI, OCV]-CharType	<See Requirement ID: "Field: Geographic">	t-2

Field: Type15-CharCount

					T
1	M	15.001-CharCount	DataLength(15.001) MO [1 to 8]		
1	M	NIEM-15.001-CharCount	Length(15.001) EQ 2		X
1	M	15.002-CharCount	DataLength(15.002) EQ 1 OR 2		B
1	M	15.003-CharCount	DataLength(15.003) EQ 2		B
1	M	15.004-CharCount	<See Requirement ID: "Field: Originating Agency".>	t-2	
1	M	15.005-CharCount	DataLength(15.005) EQ 8		B
1	M	NIEM-15.005-CharCount	DataLength(15.005) EQ 10		X
1	M	15.006-CharCount	DataLength(15.006) MO [2 to 5]		B
1	M	15.007-CharCount	DataLength(15.007) MO [2 to 5]		B
1	M	15.008-CharCount	DataLength(15.008) EQ 1		B
1	M	15.009-CharCount	DataLength(15.009) MO [1 to 5]		B
1	M	15.010-CharCount	DataLength(15.010) MO [1 to 5]		B
1	M	15.011-CharCount	DataLength(15.011) MO [3 to 5]		B
1	M	15.012-CharCount	DataLength(15.012) EQ 1 OR 2		B
1	M	15.013-CharCount	DataLength(15.013) EQ 2		B
1	O	15.016-CharCount	DataLength(15.016) MO [1 to 5]		B
1	O	15.017-CharCount	DataLength(15.017) MO [1 to 5]		B
1	O	15.020-CharCount	DataLength(15.020) MO [1 to 126]		B
1	O	15.030-CharCount	DataLength(15.030) MO [7 to 10]		B
1	O	15.903-CharCount	DataLength(15.903) MO [13 to 16]		B
1	O	15.993-	<See Requirment ID: "Field: Source Agency	t-2	

Table 73   <Table 73 specifies the Character Count for each field that contains no subfields.>

Field: Type15-Subfield CharCount

		CharCount	Name",.>						
1	O	15.996-CharCount	DataLength(15.995) EQ 64						B
1	M	15.999-CharCount	DataLength(15.999) GTE 1						B
1	O	15.018-FRAP-CharCount	ForEach(Subfield in 15.018) { Length(InfoItem:1 in Subfield) EQ 1 OR 2 }						B*
1	O	15.018-ABC-CharCount	ForEach(Subfield in 15.018) { Length(InfoItem:2 in Subfield) EQ 2 }						B*
1	O	15.024-FRMP-CharCount	ForEach(Subfield in 15.024) { Length(InfoItem:1 in Subfield)) EQ 1 OR 2 }						B*
1	O	15.024-QVU-CharCount	ForEach(Subfield in 15.024) { Length(InfoItem:2 in Subfield)) MO [1 to 3] }						B*
1	O	15.024-QAV-CharCount	ForEach(Subfield in 15.024) { Length(InfoItem:3 in Subfield)) EQ 4 }						B*
1	O	15.024-QAP-CharCount	ForEach(Subfield in 15.024) { Length(InfoItem:4 in Subfield)) MO [1 to 5] }						B*
1	O	15.902-GMT-CharCount	ForEach(Subfield in 15.902) { Length(InfoItem:1 in Subfield) EQ 15 }						T
1	O	15.902-[NAV, OWN]-CharCount	ForEach(Subfield in 15.902) { Length(InfoItem:2,3 in Subfield) MO [1 to 64] }						T
1	O	15.902-PRO-CharCount	ForEach(Subfield in 15.902) { Length(InfoItem:4 in Subfield)) MO [1 to 255] }						T

Table 73 <Table 73 specifies the Character Count for each subfield.>

1	O	NIEM-15.902-Subfield CharCount	< The treatment of subfields for validation in the XML version requires further review. Length of the first "subfield" in XML is 20.>		X*
1	O	15.904-[MAK, MOD, SER]-CharCount	Length(All(InfoItems in 15.904)) MO [1 to 50]		B*
1	O	15.995-ACN-CharCount	ForEach(Subfield in 15.995) { Length(InfoItem:1 in Subfield) MO [1 to 3] }		B*
1	O	15.995-ASP-CharCount	ForEach(Subfield in 15.995) { Length(InfoItem:2 in Subfield) EQ 1 OR 2 }		B*
1	O	15.997-SRN-CharCount	ForEach(Subfield in 15.997) { Length(InfoItem:1 in Subfield) MO [1 to 3] }		B*
1	O	15.997-RSP-CharCount	ForEach(Subfield in 15.997) { Length(InfoItem:2 in Subfield) EQ 1 OR 2 }		B*
1	O	15.998-[UTE, LTD, LTM, LTS, LGD, LGM, LGS, ELE, GDC, GCM, GCE, GCN, GRT, OSI, OCV]-CharCount	<See Requirement ID: "Field: Geographic">	t-2	
Table 73			<Table 73 specifies the Field Occurrence for each field.>		
1	-	15.[014, 015, 019, 021 to 023, 025 to 029, 031 to 199,901,905 to 994]-Occurrence	Count(15.[014, 015, 019, 021 to 023, 025 to 029, 031 to 199,901,905 to 994]) EQ 0		B
1	M	15.[001 to 013]-Occurrence	Count(15.[001 to 013]) EQ 1		B
1	-	15.[016,	Count(15.[016, 017, 018, 020, 024, 030, 902 to		B

Field: Type15-Field Occurrence

151

Field	Reference	Description	Min	Occ	Field ID	Requirement	Type	Code
					017, 018, 020, 024, 030, 902 to 904, 993, 995 to 999]- Occurrence	904, 993, 995 to 999)] LTE 1		
Field: 15.001-Record Header Value	8.15.1, Table 73, 7.1	Field 15.001 Record header. In Traditional encoding, this field contains the record length in bytes (including all information separators)		M	15.001-Record Header	<See Requirement ID "Field: xx.001-Record Header">	t-2	
	8.15.1, C.9.13	The XML name for the Type-15 record is <itl:PackagePalmprintImageRecord>, and its <biom:RecordCategoryCode> element shall have a value of "15".	1	M	NIEM-15.001-Value	ForEach(XElm(itl:PackagePalmprintImageRecord) {XElm(biom:RecordCategoryCode) EQ ASCII(15)}	t-2	X
Field: 15.002-Information Designation Character Value	8.15.2, Table 73, 7.3.1	This mandatory field shall be the IDC assigned to this Type-15 record as listed in the information item IDC for this record in Field 1.003 Transaction content/CNT.		M	15.002-IDC	<See Requirement IDs "Field: xx.002-IDC "and "Field: 1.003-Transaction Content Subfield 2 IDC Matches" >	t-2	
Field: 15.003-Impression Type Value	8.15.3, Table 73, 7.7.4.1, Table 7	This mandatory field shall indicate the manner by which the latent print was obtained. See Section 7.7.4.1 for details. <Table 73 lists the valid values for IMP.>	1	M	15.003-Value	{15.003} MO [10, 11, 28, 29]	t-2	B
Field: 15.004-Source Agency Value	8.15.4, 7.6	This is a mandatory field. See Section 7.6 for details.		M	15.004-Value	<See Requirement ID: "Field: Originating Agency".>	t-2	
Field: 15.005-Palmprint Capture Date Value	8.15.5, 7.7.2.3	This mandatory field shall contain the date that the fingerprint data contained in the record was captured.	1	M	15.005-Value	{15.005} MO [ValidLocalDate]	t-6	T
			1	M	NIEM-15.005-Value	ForEach(XElm(itl:PackagePalmprintRecord)) {XElm(nc:Date) in XElm(biom:CaptureDate)} MO [NIEM-ValidLocalDate]	t-6	X
Field: 15.006-Horizontal Line Length Value	8.15.6, Table 73, 7.7.8.1	The maximum horizontal size is limited to 65,534 pixels in Record Types-4 and 8, and to 99,999 for other record types. The minimum value is 10 pixels.		M	15.006-Value	<See Requirement ID "Field: Image HLL Value" >	t-2	
Field:	8.15.6,	<The HLL is verified by checking the image	2	M	15.006-	IF {15.011} EQ ASCII(JPEGB) OR ASCII(JPEGL)	t-11	B

Field	Reference	Description	Level	M/B	Value	Requirement	Test	
15.006-Horizontal Line Length Metadata	Table 73, 7.7.8.1	metadata if compression is used.>			Matches Image Metadata	THEN {15.006} EQ {ImageWidth-JPEGB,JPEGL} ELSE IF {15.011} EQ ASCII(JP2) OR ASCII(JP2L) THEN {15.006} EQ {ImageWidth-JP2,JP2L} ELSE IF {15.011} EQ ASCII(PNG) THEN {15.006} EQ {ImageWidth-PNG} ELSE IF {15.011} EQ ASCII(WSQ20) THEN {15.006} EQ {ImageWidth-WSQ}	t-2	
Field: 15.007-Vertical Line Length Value	8.15.7, Table 73, 7.7.8.2	The maximum vertical size is limited to 65,534 pixels in Record Types-4 and 8, and to 99,999 for other record types. The minimum value is 10 pixels.		M	15.007-Value	<See Requirement ID "Field: Image VLL Value">	t-2	B
Field: 15.007-Vertical Line Length Metadata	8.15.7, Table 73, 7.7.8.2	<The VLL is verified by checking the image metadata if compression is used.>	2	M	15.007-Matches Image Metadata	IF {15.011} EQ ASCII(JPEGB) OR ASCII(JPEGL) THEN {15.007} EQ {ImageHeight-JPEGB,JPEGL} ELSE IF {15.011} EQ ASCII(JP2) OR ASCII(JP2L) THEN {15.007} EQ {ImageHeight-JP2,JP2L} ELSE IF {15.011} EQ ASCII(PNG) THEN {15.007} EQ {ImageHeight-PNG} ELSE IF {15.011} EQ ASCII(WSQ20) THEN {15.007} EQ {ImageHeight-WSQ}	t-11	
Field: 15.008-Scale Units Value	8.15.8, Table 73, 7.7.8.3	<Table 73 lists the value constraints for SLC>		M	15.008-Value	<See Requirement ID "Field: Image SLC Value">	t-2	B
Field: 15.008-Scale Units Metadata	8.15.8, Table 73, 7.7.8.3	A value of "1" shall indicate pixels per inch. A value of "2" shall indicate pixels per centimeter. A value of "0" in this field indicates that no scale is provided, and the quotient of THPS/TVPS shall provide the pixel aspect ratio. <The SLC is verified by checking the image metadata if compression is used.>	2	M	15.008-Matches Image Metadata	IF {15.011} EQ ASCII(JPEGB) OR ASCII(JPEGL) THEN {15.008} EQ {SamplingUnits-JPEGB,JPEGL} ELSE IF {15.011} EQ ASCII(JP2) OR ASCII(JP2L) OR ASCII(WSQ20) THEN <Provide Warning "Not Tested"> ELSE IF {15.011} EQ ASCII(PNG) THEN IF {15.008} EQ 1 OR 2 THEN {SamplingUnits-PNG} EQ 1, ELSE IF {15.008} EQ 0 THEN {SamplingUnits-PNG} EQ 0	t-11	
Field: 15.009-Transmitted Horizontal Pixel Scale Value	8.15.9, Table 73, 7.7.8.4	<Table 73 lists the value constraints for THPS.>		M	15.009-Value	<See Requirement ID "Field: Image THPS Value">	t-2	B
Field: 15.009-Transmitted Horizontal	8.15.9, Table 73, 7.7.8.4	This is the integer pixel density used in the horizontal direction of the image if SLC has a value of "1" or "2". If SLC has a value of "0", this information item shall	2	M	15.009-Matches Image Metadata	IF {15.011} EQ ASCII(JPEGB) OR ASCII(JPEGL) AND {15.008} EQ 1 OR 2 THEN {HorizontalDensity-JPEGB,JPEGL}	t-11, t-12	

Field	Ref	Description		Sub-ID	Condition	Test	Type
Pixel Scale Metadata		contain the horizontal component of the pixel aspect ratio, up to 5 digits. <The THPS is verified by checking the image metadata if compression is used >			ELSE IF {15.011} EQ ASCII(JP2) OR ASCII(JP2L) OR ASCII(WSQ20) THEN <Provide Warning "Not Tested"> ELSE IF {15.011} EQ ASCII(PNG) AND {15.008} EQ 1 THEN {15.009} EQ {HorizontalDensity-PNG} * 0.0254 (meters/inch), ELSE IF {15.011} EQ ASCII(PNG) AND {15.008} EQ 2 THEN {15.009} EQ {HorizontalDensity-PNG} * 0.01 (meters/cm)	t-11	B
			2	15.009-Aspect Ratio Matches Image Metadata	M	IF {15.011} EQ ASCII(JPEGB) OR ASCII(JPEGL) AND {15.008} NEQ 1 OR 2 THEN {15.009}/{15.010} EQ {HorizontalDensity-JPEGB,JPEGL} / {VerticalDensity-JPEGB,JPEGL} ELSE IF {15.011} EQ ASCII(JP2) OR ASCII(JP2L) OR ASCII(WSQ20) THEN <Provide Warning "Not Tested"> ELSE IF {15.011} EQ ASCII(PNG) AND {15.008} NEQ 1 OR 2 THEN {15.009}/{15.010} EQ {HorizontalDensity-PNG} / {VerticalDensity-PNG}		
Field: 15.010- Transmitted Vertical Pixel Scale Value	8.15.10, Table 73, 7.7.8.5	<Table 73 lists the value constraints for TVPS.>	2	15.010-Value	M	<See Requirement ID "Field: Image TVPS Value">	t-2	
Field: 15.010- Transmitted Vertical Pixel Scale Metadata	8.15.10, Table 73, 7.7.8.5	This is the integer pixel density used in the Vertical direction of the image if SLC has a value of "1" or "2". If SLC has a value of "0", this information item shall contain the Vertical component of the pixel aspect ratio, up to 5 digits. <The TVPS is verified by checking the image metadata if compression is used >	2	15.010-Matches Image Metadata	M	IF {15.011} EQ ASCII(JPEGB) OR ASCII(JPEGL) AND {15.008} EQ 1 OR 2 THEN {15.010} EQ {VerticalDensity-JPEGB,JPEGL} ELSE IF {15.011} EQ ASCII(JP2) OR ASCII(JP2L) OR ASCII(WSQ20) THEN <Provide Warning "Not Tested"> ELSE IF {15.011} EQ ASCII(PNG) AND {15.008} EQ 1 THEN {15.010} EQ {VerticalDensity-PNG} * 0.0254 (meters/inch), ELSE IF {15.011} EQ ASCII(PNG) AND {15.008} EQ 2 THEN {15.010} EQ {VerticalDensity-PNG} * 0.01 (meters/cm)	t-11, t-12	B
			2	15.010-	M	IF {15.011} EQ ASCII(JPEGB) OR ASCII(JPEGL) AND	t-11	B

Field	Reference	Description	#	M	Requirement	Condition	Code	
					Aspect Ratio Matches Image Metadata	{15.008} NEQ 1 OR 2 THEN {15.009}/{15.010} EQ {HorizontalDensity-JPEGB,JPEGL} / {VerticalDensity-JPEGB,JPEGL} ELSE IF {15.011} EQ ASCII(JP2) OR ASCII(JP2L) OR ASCII(WSQ20) THEN <Provide Warning "Not Tested"> ELSE IF {15.011} EQ ASCII(PNG) AND {15.008} NEQ 1 OR 2 THEN {15.009}/{15.010} EQ {HorizontalDensity-PNG} / {VerticalDensity-PNG}		B
Field: 15.011-Compression Algorithm Value	8.15.11, Table 73, 7.7.9.1	For each of these fields, the entry corresponds to the appropriate *Label* entry in Table 15: Field 15.011: Compression algorithm / CGA.	1	M	15.011-Value	{15.011} MO [ASCII[NONE, JPEGB, JPEGL, JPEG, JP2, JP2L, PNG, WSQ]]		
Field: 15.011-Compression Algorithm Metadata	8.15.11, Table 73	<The CGA is verified by checking the image metadata for the compression type signature if compression is used.>	2	M	15.011-Matches Image Metadata	IF {15.011} EQ ASCII(JPEGB) OR ASCII(JPEGL) THEN Present(SOI-JPEG,JPEGL) ELSE IF {15.011} EQ ASCII(JP2) OR ASCII(JP2L) THEN Present(SigBox) ELSE IF {15.011} EQ ASCII(PNG) THEN Present(PNGSig) ELSE IF {15.011} EQ ASCII(WSQ20) THEN Present(SOI-WSQ)	t-11	B
Field: 15.012-Bits Per Pixel Value	8.15.12, Table 73, 7.7.8.6	This field shall contain an entry of "0" to "255". Any entry in this field greater than "8" shall represent a grayscale pixel with increased proportion.		M	15.012-Value	<See Requirement ID "Field: Image BPX Value">	t-2	
Field: 15.012-Bits Per Pixel Metadata	8.15.12, Table 73	<The BPX is verified by checking the image metadata for the compression type signature if compression is used.>	2	M	15.012-Matches Image Metadata	IF {15.011} EQ ASCII(JPEGB) OR ASCII(JPEGL) THEN {15.012} EQ {BPX-JPEG, JPEGL} ELSE IF {15.011} EQ ASCII(JP2) OR ASCII(JP2L) THEN 15.012} EQ {BPX-JP2,JP2L} ELSE IF {15.011} EQ ASCII(PNG) THEN {15.012} EQ {BPX-PNG} ELSE <Provide Warning "Not Tested">	t-11	B
Field: 15.013-Friction Ridge Generalized Position Value	8.15.13, Table 73, 7.7.4.2, Table 6	See Section 7.7.4.2 for details.	1	M	15.013-Value	{15.013} MO [20 to 38, 81 to 84]		B

Field	Reference	Description		Occ	Field ID	Requirement	Type	CondCode
Field: 15.014, 15.015-Reserved	Table 73	Reserved for future use only by ANSI/NIST-ITL.		-	15.014, 15.015-Reserved	<See Requirement ID: "Field: Type15 CondCode>.	t-2	
Field: 15.016-Scanned Horizontal Pixel Scale Value	8.15.14, Table 73, 7.7.8.7	See section 7.7.8 for details.		O	15.016-SHPS Value	<See Requirement IDs: "Field: Image SHPS Value">	t-2	
Field: 15.017-Scanned Vertical Pixel Scale Value	8.15.15, Table 73, 7.7.8.8	See section 7.7.8 for details.		O	15.016-Value	<See Requirement IDs: "Field: Image SVPS Value">	t-2	
Field: 15.018-Amputated or Bandaged Value	8.15.16, Table 8, Table 72	The first item is ... between 21 and 38 or 81 through 84 as chosen from Table 8. Table 72 is a list of allowable indicators for the AMPCD.	1	M ⇑	15.018-FRAP-Value	ForEach(Subfield in 15.018) { [InfoItem:1 in Subfield] MO [21 to 38, 81 to 84] }	t-2	B*
			1	M ⇑	15.018-ABC-Value	ForEach(Subfield in 15.018) { [InfoItem:2 in Subfield] MO [ASCII(XX, UP)] }	t-2	B*
Field: 15.019-Reserved	Table 73	Reserved for future use only by ANSI/NIST-ITL.		-	15.019 Reserved	<See Requirement ID: "Field: Type15 CondCode>.	t-2	
Field: 15.020-Comment Value	8.15.16, Table 73, 7.4.4	See section 7.4.4 for details.		O	15.020-Value	<See Requirement ID: "Field: Comment">.	t-2	
Field: 15.021 to 15.023-Reserved	Table 73	Reserved for future use only by ANSI/NIST-ITL.		-	15.021 to 15.023 Reserved	<See Requirement ID: "Field: Type15 CondCode>.	t-2	
Field: 15.024-Palm Quality Metric Value	8.15.17, Table 73, Table 6	<Table 73 lists the value constraints for PQM.>		O	15.024-[FRMP, QVU, QAV, QAP]-Value	<See Requirement IDs: "Field: Sample Quality Occurrences", to "Field: Sample Quality Additional Subfield" >	t-2	
Field: 15.025 to 15.029-Reserved	Table 73	Reserved for future use only by ANSI/NIST-ITL.		-	15.025 to 15.029 Reserved	<See Requirement ID: "Field: Type15 CondCode>.	t-2	
Field: 15.030-Device Monitoring	8.15.27, Table 73	<Table 73 lists the value constraints for DMM.>	1	O	15.030-Value	{15.030} MO ASCII(CONTROLLED, ASSISTED, OBSERVED, UNATTENDED, UNKNOWN)		B

Mode Value							
Field: 15.031 to 15.199- Reserved	Table 73	Reserved for future use only by ANSI/NIST-ITL.	-	15.031 to 15.199 Reserved	<See Requirement ID: "Field: Type15-CondCode>.	t-2	
Field: 15.200 to 15.900- User Defined	Table 73	User Defined Fields	-	15.200 to 15.900- User Defined	TRUE		B
Field: 15.901- Reserved	Table 73	Reserved for future use only by ANSI/NIST-ITL.	-	15.901- Reserved	<See Requirement ID: "Field: Type15-CondCode>.	t-2	
Field: 15.902- Annotated Information Value	8.15.20, Table 73	This is an optional field, listing the operations performed on the original source in order to prepare it for inclusion in a biometric record type. See Section 7.4.1.	O	15.902- [GMT, NAV, OWN, PRO]-Value	<See Requirement ID: "Field: xx.902-ANN" >.	t-2	
Field: 15.903- Device Unique Identifier Value	8.15.31, Table 73	This is an optional field. See Section 7.7.1.1.	O	15.903- Value	<See Requirement ID: "Field: Device ID" >.	t-2	
Field: 15.904- Make/Model/Serial Number Value	8.15.32, Table 73	This is an optional field. See Section 7.7.1.2.	O	15.904- [MAK, MOD, SER]-Value	<See Requirement ID: "Field: Make Model" >.	t-2	
Field: 15.905 to 15.992- Reserved	Table 73	Reserved for future use only by ANSI/NIST-ITL.	-	15.905, 15.992- Reserved	<See Requirement ID: "Field: Type15-CondCode>.	t-2	
Field: 15.993- Source Agency Name	8.15.24, Table 73	This is an optional field. It may contain up to 125 Unicode characters.	O	15.993- Value	<See Requirment ID: " Field: Source Agency Name" .>	t-2	
Field: 15.994- Reserved	Table 73	Reserved for future use only by ANSI/NIST-ITL.	-	15.994- Reserved	<See Requirement ID: "Field: Type15-CondCode>.	t-2	
Field: 15.995- Associated Context Value	8.15.33, Table 73	See Section 7.3.3	O	15.995- [ACN, ASP]-Value	<See Requirement IDs: "Field: xx.995-ASC-ACN" and "Field: xx.995-ASC-ASP">.	t-2	
Field: 15.996-	8.15.34, Table 73	See Section 7.5.2	O	15.996- Value	<See Requirement ID: "Field: HAS">	t-2	

Hash Value							
Field: 15.997- Source Representation Value	8.15.35, Table 73	See Section 7.3.2	O	15.997-[SRN, RSP]-Value	\<See Requirement IDs: "Field: xx.997-SOR" and "Field: xx.997-SOR-SRN" and "Field: xx.997-SOR-RSP"\>.	t-2	B
Field: 15.998- Geographic Sample Acquisition Location Value	8.15.36, Table 73	See Section 7.7.3	O	15.998-[UTE, LTD, LTM, LTS, LGD, LGM, LGS, ELE, GDC, GCM, GCE, GCN, GRT, OSI, OCV]-Value	\<See Requirement IDs: "Field: Geographic", "Field: Geographic-Subfield 1" through "Field: Geographic-Values-SubField 15"\>.	t-2	B
Field: 15.999- Image Data Valid	8.15.37, Table 73	This field contains the palmprint image. \<The image metadata is checked for validity.\>	2 D	15.999- Uncompressed Image Length	IF {15.011} EQ ASCII(NONE) THEN Length(15.999) EQ {15.006} * {15.007}		
			2 D	15.999- Valid Image Format	IF {15.011} EQ ASCII(JPEGB) OR ASCII(JPEGL) THEN Present(JFIF, SOI-JPEGB, JPEGL, SOF-JPEGB, JPEGL, EOI-JPEG, JPEGL) ELSE IF {15.011} EQ ASCII(JP2) OR ASCII(JP2L) THEN Present(SigBox, HeadBox, ImgBox, EOI-JP2, JP2L) ELSE IF {15.011} EQ ASCII(PNG) THEN Present(PNGSig, IHDR, IDAT, IEND) ELSE IF {15.999} EQ ASCII(WSQ20) THEN Present(SOI-WSQ, SOF-WSQ, SOB-WSQ, EOI-WSQ)	t-11	B
Field: 15.999- Image WSQ Version 3.1	7.7.9.1	Only version 3.1 or higher shall be used for compressing grayscale fingerprintdata at 500 ppi class with a platen area of 2 inches or greater in height. WSQ 2.0 or higher may be used for 500 ppi class data taken from a platen of less than 2 inches in height. WSQ shall not be used for other than the 500 ppi class.	2 D	15.999- Valid WSQ Encoder Version	IF {15.011} EQ ASCII(WSQ20) THEN {Encoder Version} EQ 1 OR 2	t-11	B

Table 6.13 - Assertions for Record Type 17 - Iris Image Record

Requirement ID	Reference in Base Standard	Requirement Summary	Level	Status	Assertion ID	Test Assertion	Test Note	Implementation Support	Supported Range	Test Result	Applicability
						8.17: Record Type-17: Iris image record					
Field: Type17-Subfield Occurrence	Table 75	<Table 75 specifies which fields contain subfields as well as the number of occurrences permitted >	1	M	17.[001 to 017, 019 to 023, 025 to 028, 030 to 036, 040, 041, 993, 996, 998, 999]-SubfieldCount	Count(Subfields in 17.[001 to 017, 019 to 023, 025 to 028, 030 to 036, 040, 041, 993, 996,998,999]) EQ 1					T
			1	M	17.[001 to 005, 015, 017, 019 to 023, 025 to 028, 030 to 032, 040, 041, 993, 996]-InfoItemCount	Count(InfoItems in Subfield:1 in 17.[001 to 015, 017, 019 to 023, 025 to 028, 030 to 032, 040, 041, 993, 996]) EQ 1					T
			1	O	17.016-InfoItemCount	Count(InfoItems in 17.016) EQ 3					T
			1	O	17.019-InfoItemCount	Count(InfoItems in 17.019) EQ 3					T
			1	O	17.024-SubfieldCount	Count(Subfields in 17.024) MO [1 to 9]					T
			1	O	17.024-InfoItemCount	ForEach(Subfield in 17.024) { Count(InfoItems in Subfield) EQ 3					T

1	O	17.027-InfoItemCount	} Count(InfoItems in 17.027) EQ 2	T
2	O	17.033-InfoItemCount	Count(InfoItems in 17.033) EQ 2 + 2*{InfoItem:2 in 17.033}	T
2	O	17.034-InfoItemCount	Count(InfoItems in 17.034) EQ 2 + 2*{InfoItem:2 in 17.034}	T
2	O	17.035-InfoItemCount	Count(InfoItems in 17.035) EQ 2 + 2*{InfoItem:2 in 17.035}	T
2	O	17.036-InfoItemCount	Count(InfoItems in 17.036) EQ 2 + 2*{InfoItem:2 in 17.036}	T
2	O	17.037-SubfieldCount	Count(Subfields in 17.037) GTE 1	T
2	O	17.037-InfoItemCount	Foreach(Subfield in 17.037) { Count(InfoItems in Subfield) EQ 3 + 2*{InfoItem:3 in Subfield} }	T
1	O	17.902-SubfieldCount	Count(Subfields in 17.902) GTE 1	T
1	O	17.902-InfoItemCount	ForEach(Subfield in 17.902) { Count(InfoItems in Subfield) EQ 4 }	T
1	O	17.995-SubfieldCount	Count(Subfields in 17.995) MO [1 to 255]	T
1	O	17.995-InfoItemCount	ForEach(Subfield in 17.995) { Count(InfoItems in Subfield) EQ 2 OR 3 }	T
1	O	17.997-SubfieldCount	Count(Subfields in 17.997) MO [1 to 255]	T
1	O	17.997-InfoItemCount	ForEach(Subfield in 17.997) { Count(InfoItems in Subfield) EQ 1 OR 2 }	T
	O	17.998-	<See Requirement ID: "Field: Geographic">	t-2

Field	Reference	Description			Subfields	Condition						B
Field:17-Type17-CondCode	Table 75	<Table 75 specifies the Condition Code for each field.>	1	-	[17.001 to 17.005, 17.013]-Mandatory CondCode	Present(17.001 to 17.005)						B
			1	-	[17.018, 17.029, 17.038, 17.039, 17.042 to 17.199, 17.901, 17.903 to 17.992, 17.994]-Reserved	NOT Present(17.018, 17.029, 17.038, 17.039, 17.042 to 17.199, 17.901, 17.903 to 17.992, 17.994)						B
Record: 17.006 to 17.012 Dependent	Table 75, 8.17.6 to 8.17.12	This field is mandatory if present in Field 17.999. Otherwise it is absent	2	D	[17.006 to 17.012]-CondCode Dependent	Present(17.999) IFF Present(17.006 to 17.012)						B
Field:17.015-Rotation Uncertainty Dependent	Table 75, 8.17.15	This field is mandatory if Field 17.014: Rotation angle of eye / RAE is present.	2	D	17.015-CondCode Dependent	IF Present(17.014) THEN Present(17.015)						B
Field:17.027-Specified Spectrum Dependent	Table 75, 8.17.26	This field shall only be present if Field 17.025: Effective acquisition spectrum / EAS has a value of 'DEFINED'.	2	D	17.027-CondCode Dependent	IF Present(17.027) THEN {17.025} EQ ASCII(DEFINED)						B
Field:17.999-Iris Image Dependent	Table 75, 8.17.45	This field contains the iris image. It shall contain an image, unless ...Field 17.028: Damaged or missing eye / DME is in this record...in which case DATA is optional.	2	D	17.999-CondCode Dependent	IF NOT Present(17.028) THEN Present(17.999)						B
Field:17-Type17-CharType	8.17, Table 75	<Table 75 specifies the Character Type for each field that contains no subfields.>	1	-	17.[001,002,003,006 to 010, 012, 022, 023, 026, 031, 032, 040, 041]-CharType	Bytes(17.[001,002,003,006 to 010, 012, 022, 023, 026, 031, 032, 040, 041]) MO [0x30 to 0x39]						B
			1	-	17.[013, 020, 025, 028, 030]-	Bytes(17.[013, 020, 025, 028, 030]) MO [0x20, 0x41 to 0x5A, 0x61 to 0x7A]						B

Field	Reference	Count	M/O	Sub-field	Description		Char
		1		CharType	Bytes(17.011)[0x30 to 0x39, 0x20, 0x41 to 0x5A, 0x61 to 0x7A]		B
		1	-	17.011-CharType	Bytes(17.[014, 015])[0x30 to 0x39, 0x41 to 0x46, 0x61 to 0x66]		B
		1	-	17.[014, 015]-CharType	<See Requirement ID: "Field: Originating Agency".>	t-2	T
		1	M	17.004-CharType	Bytes(17.005) MO [0x30 to 0x39]		X
		1	M	17.005-CharType	Bytes(17.005) MO [0x30 to 0x39, 0x2D]		B
		1	M	NIEM-17.005-CharType	Bytes(17.017) MO [0x20 to 0x7E]		B
		1	O	17.017-CharType	<See Requirement ID: "Field: Source Agency Name".>	t-2	B*
		1	O	17.993-CharType	Bytes(17 996) MO [0x30 to 0x39,0x41 to 0x46, 0x61 to 0x66]		B*
		1	O	17.996-CharType	Bytes(All(InfoItems in 17.016)) MO [0x30 to 0x39]		B*
Field: Type17-Subfield CharType	8.17, Table 75 <Table 75 specifies the Character Type for each subfield.>	1	O	17.016-[IHO, IVO, IST]-CharType	TRUE		B*
		1	O	17.019-[MAK, MOD, SER]-CharType	ForEach(Subfield in 17.024) { Bytes(InfoItem:1,3 in Subfield)) MO [0x30 to 0x39] }		B*
		1	O	17.024-[QVU, QAP]-CharType	ForEach(Subfield in 17.024) { Bytes(InfoItem:2 in Subfield)) MO [0x30 to 0x39,0x41 to 0x46, 0x61 to 0x66] }		B*
		1	O	17.024-QAV-CharType	Bytes(All(InfoItems in 17.027)) MO [0x30 to 0x39]		B*
		1	M	17.027-[LOW, HIG]-CharType	Bytes(InfoItem:1 in 17.033) MO [0x20,0x41 to 0x5A, 0x61 to 0x7A]		B*
		1	M	17.033-BYC-CharType	For(X EQ 2 to Count(InfoItems in 17.033)) {		B*
				17.033-[NOP, HPO,			

		VPO]-CharType	Bytes(InfoItem:X in 17.033) MO [0x30 to0x39]	B*
1	O	17.034-BYC-CharType	Bytes(InfoItem:1 in 17.034)) MO [0x20, 0x41 to 0x5A, 0x61 to0x7A]]	B*
1	O	17.034-[NOP, HPO, VPO]-CharType	For(X EQ 2 to Count(InfoItems in 17.034)) { Bytes(InfoItem:X in 17.034) MO [0x30 to0x39] }	
1	O	17.035-BYC-CharType	Bytes(InfoItem:1 in 17.035)) MO [0x20, 0x41 to 0x5A, 0x61 to0x7A]]	B*
1	O	17.035-[NOP, HPO, VPO]-CharType	For(X EQ 2 to Count(InfoItems in 17.035)) { Bytes(InfoItem:X in 17.035) MO [0x30 to0x39] }	B*
1	O	17.036-BYC-CharType	Bytes(InfoItem:1 in 17.036)) MO [0x20, 0x41 to 0x5A, 0x61 to0x7A]]	B*
1	O	17.036-[NOP, HPO, VPO]-CharType	For(X EQ 2 to Count(InfoItems in 17.036)) { Bytes(InfoItem:X in 17.036) MO [0x30 to0x39] }	B*
1	O	17.037-[OCY, OCT]-CharType	ForEach(Subfield in 17.037) { Bytes(InfoItem:1,2 in Subfield)) MO [0x20, 0x41 to 0x5A, 0x61 to 0x7A]] }	B*
1	O	17.037-[NOP, HPO. VPO]-CharType	ForEach(Subfield in 17.037) { For(X EQ 3 to Count(InfoItems in Subfield)) { Bytes(InfoItem:X in Subfield) MO [0x30 to 0x39] } }	B*
1	O	17.902-GMT-CharType	ForEach(Subfield in 17.902) { Bytes(InfoItem:1 in Subfield) MO [0x30 to 0x39,0x5A] }	T

	Cond	Field	Value		Type
1	O	17.902-[NAV, OWN, PROJ]-CharType	TRUE		T
1	O	NIEM-17.902-Subfield CharType	< The treatment of subfields for validation in the XML version requires further review. Byte values allowed for first "subfield" in XML are 0x30 to 0x39, 0x3A, 0x54, 0x5A.>		X*
1	O	17.995–[ACN, ASP]-CharType	Bytes(All(InfoItem:1,2 in 17 995)) MO [0x30 to 0x39]		B*
1	O	17.997–[SRN, RSP]-CharType	Bytes(All(InfoItem:1,2 in 17 997)) MO [0x30 to 0x39]		B*
1	O	17.998–[UTE, LTD, LTM, LTS, LGD, LGM, LGS, ELE, GDC, GCM, GCE, GCN, GRT, OSI, OCV]-CharType	<See Requirement ID: "Field: Geographic">	t-2	
1	M	17.001-CharCount	DataLength(17.001) MO [1 to 8]		T
1	M	NIEM-17.001-CharCount	Length(17.001) EQ 2		X
1	M	17.002-CharCount	DataLength(17.002) EQ 1 OR 2		B
1	M	17.003-CharCount	DataLength(17.003) EQ 1		B
1	M	17.004-CharCount	<See Requirement ID: "Field: Originating Agency".>	t-2	B
1	M	17.005-CharCount	DataLength(17.005) EQ 8		B
1	M	NIEM-17.005-CharCount	DataLength(17.005) EQ 10		X
1	M	17.006-CharCount	DataLength(17.006) MO [2 to 5]		B
1	M	17.007-CharCount	DataLength(17.007) MO [2 to 5]		B

Table 75 <Table 75 specifies the Character Count for each field that contains no subfields.>

Field: Type17-CharCount

1	M	17.008-CharCount	DataLength(17.008) EQ 1	B
1	M	17.009-CharCount	DataLength(17.009) MO [1 to 5]	B
1	M	17.010-CharCount	DataLength(17.010) MO [1 to 5]	B
1	M	17.011-CharCount	DataLength(17.011) MO [3 to 4]	B
1	M	17.012-CharCount	DataLength(17.012) EQ 1 OR 2	B
1	M	17.013-CharCount	DataLength(17.013) MO [3 to 4]	B
1	O	17.014-CharCount	DataLength(17.014) MO [1 to 4]	B
1	D	17.015-CharCount	DataLength(17.015) MO [1 to 4]	B
1	O	17.017-CharCount	DataLength(17.017) MO [13 to 16]	B
1	O	17.020-CharCount	DataLength(17.020) EQ 3	B
1	O	17.021-CharCount	DataLength(17.021) MO [1 to 126]	B
1	O	17.022-CharCount	DataLength(17.022) MO [1 to 5]	B
1	O	17.023-CharCount	DataLength(17.023) MO [1 to 5]	B
1	O	17.025-CharCount	DataLength(17.025) MO [3 to 9]	B
1	O	17.026-CharCount	DataLength(17.026) MO [2 to 4]	B
1	O	17.028-CharCount	DataLength(17.028) EQ 2	B
1	O	17.030-CharCount	DataLength(17.030) MO [7 to 10]	B
1	O	17.031-CharCount	DataLength(17.031) EQ 2	B
1	O	17.032-CharCount	DataLength(17.032) EQ 1	B
1	O	17.040-CharCount	DataLength(17.040) MO [1 to 7]	B
1	O	17.041-CharCount	DataLength(17.041) EQ 1 OR 2	B
1	O	17.993-	<See Requirment ID: "Field: Source Agency	t-2

		CharCount	Name".>	
Field: Type17-Subfield CharCount	Table 75		<Table 75 specifies the Character Count for each subfield.>	
	1	O	17.996-CharCount	DataLength(17.995) EQ 64	B
	1	M	17.999-CharCount	DataLength(17.999) GTE 1	B
	1	O	17.016-[IHO, IVO, IST]-CharCount	Length(All(InfoItems in 17.016)) EQ 1	B*
	1	O	17.019—[MAK, MOD, SER]-CharCount	Length(All(InfoItems in 17.019)) MO [1 to 50]	B*
	1	O	17.024—QVU-CharCount	ForEach(Subfield in 17.024) { Length(InfoItem:1 in Subfield) MO [1 to 3] }	B*
	1	O	17.024-QAV-CharCount	ForEach(Subfield in 17.024) { Length(InfoItem:2 in Subfield)) EQ 4 }	B*
	1	O	17.024-QAP-CharCount	ForEach(Subfield in 17.024) { Length(InfoItem:3 in Subfield)) MO [1 to 5] }	B*
	1	O	17.027-[LOW, HIG]-CharCount	Length(All(InfoItems in 17.027)) EQ 3 OR 4	B*
	1	O	17.033—BYC-CharCount	ForEach(Subfield in 17.033) { Length(InfoItem:1 in Subfield)) EQ 1 }	B*
	1	O	17.033-NOP-CharCount	ForEach(Subfield in 17.033) { Length(InfoItem:2 in Subfield)) EQ 1 OR 2 }	B*
	1	O	17.033-[HPO, VPO]-CharCount	ForEach(Subfield in 17.033) { ForEach(InfoItem in Subfield ST InfoItem NOT InfoItem:1 OR InfoItem 2 in Subfield) {	B*

166

			Length(InfoItem) MO [1 to 5] }
1	O	17.034-BYC-CharCount	ForEach(Subfield in 17.034) { Length(InfoItem:1 in Subfield)) EQ 1	B*
1	O	17.034-NOP-CharCount	ForEach(Subfield in 17.034) { Length(InfoItem:2 in Subfield)) EQ 1 OR 2	B*

1	O	17.034-[HPO, VPO]-CharCount	ForEach(Subfield in 17.034)
			{
			ForEach(InfoItem in Subfield ST InfoItem NOT InfoItem:1 OR InfoItem 2 in Subfield)
			{
			Length(InfoItem) MO [1 to 5]
			}
				B*

1	O	17.035-BYC-CharCount	ForEach(Subfield in 17.035) { Length(InfoItem:1 in Subfield)) EQ 1	B*
1	O	17.035-NOP-CharCount	ForEach(Subfield in 17.035) { Length(InfoItem:2 in Subfield)) EQ 1 OR 2	B*

1	O	17.035-[HPO, VPO]-CharCount	ForEach(Subfield in 17.035)
			{
			ForEach(InfoItem in Subfield ST InfoItem NOT InfoItem:1 OR InfoItem 2 in Subfield)
			{
			Length(InfoItem) MO [1 to 5]
			}
			}
				B*

1	O	17.036-BYC-CharCount	ForEach(Subfield in 17.036) { Length(InfoItem:1 in Subfield)) EQ 1	B*
1	O	17.036-NOP-CharCount	ForEach(Subfield in 17.036) { Length(InfoItem:2 in Subfield)) EQ 1 OR 2	B*
1	O	17.036-[HPO, VPO]-	ForEach(Subfield in 17.036) { ForEach(InfoItem in Subfield ST InfoItem NOT	B*

		CharCount	InfoItem::1 OR InfoItem 2 in Subfield)
			 {
			 Length(InfoItem) MO [1 to 5]
			 }
				B*

1	O	17.037-[OCV, OCT]-CharCount	ForEach(Subfield in 17.037)
			 {
			 Length(InfoItem:1,2 in Subfield)) EQ 1 AND
			 }
				B*

1	O	17.037-NOP]-CharCount	ForEach(Subfield in 17.037)
			 {
			 Length(InfoItem:3 in Subfield)) MO [1 to 2]
			 }
				B*

1	O	17.037-[HPO, VPO]-CharCount	ForEach(Subfield in 17.037)
			 ForEach(InfoItem in Subfield ST InfoItem NOT InfoItem::1 OR InfoItem 2 in Subfield)
			 {
			 Length(InfoItem) MO [1 to 5]
			 }
				B*

1	O	17.902-GMT-CharCount	ForEach(Subfield in 17.902)
			 {
			 Length(InfoItem:1 in Subfield) EQ 15
			 }
				T

1	O	17.902-[NAV, OWN]-CharCount	ForEach(Subfield in 17.902)
			 {
			 Length(InfoItem:2,3 in Subfield) MO [1 to 64]
			 }
				T

1	O	17.902-PRO-CharCount	ForEach(Subfield in 17.902)
			 {
			 Length(InfoItem:4 in Subfield)) MO [1 to 255]
			 }
				T

1	O	NIEM-17.902-Subfield CharCount	< The treatment of subfields for validation in the XML version requires further review. Length of the first "subfield" in XML is 20.>	X*

1	O	17.995-ACN-CharCount	ForEach(Subfield in 17.995)
			 {
			 Length(InfoItem:1 in Subfield) EQ 1 OR 2
			 }
				B*

1	O	17.995-ASP-CharCount	ForEach(Subfield in 17.995)
			 {
			 Length(InfoItem:2 in Subfield) MO [1 to 3]
			 }
				B*

Field	Reference	Description			Field ID	Condition	t-2	Status
			1	O	17.997-SRN-CharCount	ForEach(Subfield in 17.997) { Length(InfoItem:1 in Subfield) MO [1 to 3] }		B*
			1	O	17.997-RSP-CharCount	ForEach(Subfield in 17.997) { Length(InfoItem:2 in Subfield) EQ 1 OR 2 }		B*
Field: Type17-Field Occurrence	Table 75	<Table 75 specifies the Field Occurrence for each field.>	1	O	17.998-[UTE, LTD, LTM, LTS, LGD, LGM, LGS, ELE, GDC, GCM, GCE, GCN, GRT, OSI, OCV]-CharCount	<See Requirement ID: "Field: Geographic">	t-2	
			1	-	17.[018, 029, 038, 039, 042 to 199, 901, 903 to 992, 994]-Occurrence	Count(17.[018, 029, 038, 039, 042 to 199, 901, 903 to 992, 994]) EQ 0		B
			1	M	17.[001 to 013, 999]-Occurrence	Count(17.[001 to 005]) EQ 1		B
			1	-	17.[014 to 017, 019 to 024 to 028, 030, 031 to 037, 040, 041, 902, 993, 995 to 998]-Occurrence	Count(17.[006 to 017, 019 to 028, 030 to 037, 040, 041, 902, 993, 995 to 999]) LTE 1		B
Field: 17.001-Record Header	8.17.1, Table 75, 7.1	Field 17.001 Record header. In Traditional encoding, this field contains the record length in bytes (including all information separators)	1	M	17.001-Record Header	<See Requirement ID "Field: xx.001-Record Header">	t-2	
Value	8.17.1, C.9.15	The XML name for the Type-17 record is <itl:PackageIrisImageRecord>, and its <biom:RecordCategoryCode> element shall have a value of "17".	1	M	NIEM-17.001-Value	ForEach(XElm(itl:PackageIrisImageRecord) {XElm(biom:RecordCategoryCode) EQ ASCII(17) }		X
Field:	8.17.2,	This mandatory field shall contain the IDC		M	17.002-IDC	<See Requirement IDs "Field: xx.002-IDC" and	t-2	

Reference	Field	Description			Field ID	Requirement		
		assigned to this Type-17 record as listed in the information item IDC for this record in Field 1.003 Transaction content/CNT.				"Field: 1.003-Transaction Content Subfield 2 IDC Matches" >		
Table 75, 7.3.1	17.002- Information Designation Character Value							
8.17.3, Table 75, 7.7.4.1	Field: 17.003-Eye Label Value	<Table 75 lists the valid values for ELR.>	1	M	17.003- Value	{17.003} MO [0 to 2]		B
8.17.4, 7.6	Field: 17.004- Source Agency Value	This is a mandatory field. See Section 7.6 for details.		M	17.004- Value	<See Requirement ID: "Field: Originating Agency".>	t-2	
8.17.5, 7.7.2.3	Field: 17.005-Iris Capture Date Value	This mandatory field shall contain the date that the iris biometric data contained in the record was captured.	1	M	17.005- Value	{17.005} MO [ValidLocalDate]	t-6	T
				M	NIEM- 17.005- Value	ForEach(XElm(iti:PackageIrisImageRecord)) { XElm(nc:Date) in XElm(biom:CaptureDate)} MO [NIEM-ValidLocalDate] }	t-6	X
8.17.6, Table 75, 7.7.8.1	Field: 17.006- Horizontal Line Length Value	The maximum horizontal size is limited to 65,534 pixels in Record Types-4 and 8, and to 99,999 for other record types. The minimum value is 10 pixels.	2	M	17.006- Value	<See Requirement ID "Field: Image HLL Value" >	t-2	
8.17.6, Table 75, 7.7.8.1	Field: 17.006- Horizontal Line Length Metadata	<The HLL is verified by checking the image metadata if compression is used.>		M	17.006- Matches Image Metadata	IF {17.011} EQ ASCII(JP2) OR ASCII(JP2L) THEN {17.006} EQ {ImageWidth-JP2,JP2L} ELSE IF {17.011} EQ ASCII(PNG) THEN {17.006} EQ {ImageWidth-PNG}	t-11	B
8.17.7, Table 75, 7.7.8.2	Field: 17.007- Vertical Line Length Value	The maximum vertical size is limited to 65,534 pixels in Record Types-4 and 8, and to 99,999 for other record types. The minimum value is 10 pixels.	2	M	17.007- Value	<See Requirement ID "Field: Image VLL Value" >	t-2	
8.17.7, Table 75, 7.7.8.2	Field: 17.007- Vertical Line Length Metadata	<The VLL is verified by checking the image metadata if compression is used.>		M	17.007- Matches Image Metadata	IF {17.011} EQ ASCII(JP2) OR ASCII(JP2L) THEN {17.007} EQ {ImageHeight-JP2,JP2L} ELSE IF {17.011} EQ ASCII(PNG) THEN {17.007} EQ {ImageHeight-PNG}	t-11	B
8.17.8, Table 75, 7.7.8.3	Field: 17.008-Scale Units Value	<Table 75 lists the value constraints for SLC>		M	17.008- Value	<See Requirement ID "Field: Image SLC Value" >	t-2	
8.17.8, Table 75,	Field: 17.008-	A value of "1" shall indicate pixels per inch.			17.008- Matches	IF {17.011} EQ ASCII(JP2) OR ASCII(JP2L) THEN <Provide Warning "Not Tested">		B

Field	Reference	Description		Requirement	Logic	Test	
Scale Units Metadata	7.7.8.3	A value of "2" shall indicate pixels per centimeter. A value of "0" in this field indicates that no scale is provided, and the quotient of THPS/TVPS shall provide the pixel aspect ratio. <The SLC is verified by checking the image metadata if compression is used.>		Image Metadata	ELSE IF {17.011} EQ ASCII(PNG) THEN IF {17.008} EQ 1 OR 2 THEN {SamplingUnits-PNG} EQ 1, ELSE IF {17.008} EQ 0 THEN { SamplingUnits-PNG} EQ0		
Field: 17.009-Transmitted Horizontal Pixel Scale Value	8.17.9, Table 75, 7.7.8.4	<Table 75 lists the value constraints for THPS.>	M	17.009-Value	<See Requirement ID "Field: Image THPS Value">	t-2	
Field: 17.009-Transmitted Horizontal Pixel Scale Metadata	8.17.9, Table 75, 7.7.8.4	This is the integer pixel density used in the horizontal direction of the image if SLC has a value of "1" or "2". If SLC has a value of "0", this information item shall contain the horizontal component of the pixel aspect ratio, up to 5 digits. <The THPS is verified by checking the image metadata if compression is used >	B	17.009-Matches Image Metadata	IF {17.011} EQ ASCII(JP2) OR ASCII(JP2L) THEN <Provide Warning "Not Tested"> ELSE IF {17.011} EQ ASCII(PNG) AND {17.008} EQ 1 THEN {17.009} EQ {HorizontalDensity-PNG} * 0.0254 (meters/inch) ELSE IF {17.011} EQ ASCII(PNG) AND {17.008} EQ 2 THEN {17.009} EQ {HorizontalDensity-PNG} * 0.01 (meters/cm)	t-11, t-12	B
				17.009-Aspect Ratio Matches Image Metadata	IF {17.011} EQ ASCII(JP2) OR ASCII(JP2L) THEN <Provide Warning "Not Tested"> ELSE IF {17.011} EQ ASCII(PNG) {17.008} NEQ 1 OR 2 THEN {17.009}/{17.010} EQ {HorizontalDensity-PNG} / {VerticalDensity-PNG}		B
Field: 17.010-Transmitted Vertical Pixel Scale Value	8.17.10, Table 75, 7.7.8.5	<Table 75 lists the value constraints for TVPS.>	M	17.010-Value	<See Requirement ID "Field: Image TVPS Value">	t-2	
Field: 17.010-Transmitted Vertical Pixel Scale Metadata	8.17.10, Table 75, 7.7.8.5	This is the integer pixel density used in the Vertical direction of the image if SLC has a value of "1" or "2". If SLC has a value of "0", this information item shall contain the Vertical component of the pixel aspect ratio, up to 5 digits. <The TVPS is verified by checking the image metadata if compression is used >	B	17.010-Matches Image Metadata	IF {17.011} EQ ASCII(JP2) OR ASCII(JP2L) THEN <Provide Warning "Not Tested"> ELSE IF {17.011} EQ ASCII(PNG) AND {17.008} EQ 1 THEN {17.010} EQ {VerticalDensity-PNG} * 0.0254 (meters/inch), ELSE IF {17.011} EQ ASCII(PNG) AND {17.008} EQ 2	t-11, t-12	B

Field	Reference	Description	#	M/O/D	Requirement ID / Value	Condition	Code	
					17.010-Aspect Ratio Matches Image Metadata	THEN {17.010} EQ {VerticalDensity-PNG} * 0.01 (meters/cm)		B
Field: 17.011-Compression Algorithm Value	8.17.11, Table 75, 7.7.9.1	For each of these fields, the entry corresponds to the appropriate *Label* entry in Table 15: Field 17.011: Compression algorithm / CGA.		M	17.011-Value	<See Requirement ID "Field: Type17 Compression".>	t-2	
Field: 17.011-Compression Algorithm Metadata	8.17.11, Table 75	<The CGA is verified by checking the image metadata for the compression type signature if compression is used.>	2	M	17.011-Matches Image Meta Data	IF {17.011} EQ ASCII(JP2) OR ASCII(JP2L) THEN Present(SigBox) ELSE IF {17.011} EQ ASCII(PNG) THEN Present(PNGSig)	t-11	B
Field: 17.012-Bits Per Pixel Value	8.17.12, Table 75, 7.7.8.6	This field shall contain an entry of "8" for normal grayscale values of "0" to "255". Any entry in this field greater than "8" shall represent a grayscale pixel with increased proportion.	2	M	17.012-Value	<See Requirement ID "Field: Image BPX Value" >	t-2	
Field: 17.012-Bits Per Pixel Metadata	8.17.12, Table 75	<The BPX is verified by checking the image metadata for the compression type signature if compression is used.>	2	M	17.012-Matches Image Metadata	IF {17.011} EQ ASCII(JP2) OR ASCII(JP2L) THEN {17.012} EQ {BPX-JP2,JP2L} ELSE IF {17.011} EQ ASCII(PNG) THEN {17.012} EQ {BPX-PNG}	t-11	B
Field: 17.013-Color Space Value	8.17.13, Table 75, 7.7.10	Table 16 lists the codes and their descriptions for each of the available color spaces used within this standard. All other color spaces are to be marked as undefined.	1	M	17.013-Value	<See Requirement ID: "Field: Image CSP Value".>	t-2	
Field: 17.014-Rotation Angle of Eye Value	8.17.14, Table 75	The in-plane eye rotation angle shall be recorded as angle = round (65535 * angle / 360) modulo 65535. The value "FFFF" indicates that rotation angle of eye is undefined. <Table 75 lists the value constraints for RAE.>	1	O	17.014-Value	{17.014} MO [0x0000 to 0xFFFF]		B
Field: 17.015-Rotation	8.17.15, Table 75	The rotation uncertainty is non-negative and equal to [round (65535* uncertainty / 180)]. The uncertainty is measured in	1	D	17.015-Value	{17.015} MO [0x0000 to 0xFFFF]		B

Field	Reference	Status	Mnemonic	Count	Description	Requirement / Value	Note	B*
Uncertainty Value					degrees and is the absolute value of maximum error. The value "FFF" indicates that uncertainty is undefined. <Table 75 lists the value constraints for RAU.>			
Field: 17.015-Rotation Uncertainty Conditional	8.17.15, Table 75	D	17.015-Conditional		This optional field shall indicate the uncertainty in the in-plane eye rotation given in Field 17.014: Rotation angle of eye / RAE. This field is mandatory if Field 17.014: Rotation angle of eye / RAE is present.	<See Requirement ID: "Field: 17.015-Rotation Uncertainty Dependent" >	t-2	B*
Field: 17.016-Image Property Code Value	8.17.16, Table 75	O	17.016-[IHO, IVO]-Value	1	<Table 75 lists the value constraints for IPC.>	{InfoItem:1,2 in 17.016} MO [0 to 2] AMD MO [Integers]		B*
		O	17.016-IST-Value	1		{InfoItem:3 in 17.016} EQ 0 OR 1		
Field: 17.017-Device Unique Value	8.17.17, Table 75	O	17.017-Value		See Section 7.7.1.1 for details. <Table 75 lists the value constraints for DUI.>	<See Requirement ID: "Field: Device ID" >	t-2	
Field: 17.018-Deprecated	Table 75	-	17.018-Deprecated		Deprecated; See ANSI/NIST-ITL 1-2007 for a description of this field. Not to be used for any new transaction.	<See Requirement ID: "Field: Type17-CondCode">.	t-2	
Field: 17.019-Make/Model/Serial Number Value	8.17.18, 7.7.1.2, Table 75	O	17.019-[MAK, MOD, SER]-Value		See Section 7.7.1.2 for details. <Table 75 lists the value constraints for MMS.>	<See Requirement ID: "Field: Make Model" .>	t-2	
Field: 17.020-Eye Color Value	8.17.19, 7.7.11, Table 75, Table 17	O	17.020-Value		See Section 7.7.11 and Table 17 for details on entering values to this field. <Table 75 lists the value constraints for ECL.>	<See Requirement ID: "Field: Image ECL Value" .>	t-2	
Field: 17.021-Comment Value	8.17.20, Table 75, 7.4.4	O	17.021-Value		See section 7.4.4 for details.	<See Requirement ID: "Field: Comment>.	t-2	
Field: 17.022-Scanned Horizontal Pixel Scale Value	8.17.21, 7.7.8.7, 7.4.4	O	17.022-Value		See section 7.7.8.7 for details.	<See Requirement ID: "Field: Image SHPS Value".>	t-2	

Field	Reference	Value Constraints	Min	M/O	Field Code	Requirement	Type	Cond
Field: 17.023- Scanned Vertical Pixel Scale Value	8.17.22, 7.7.8.8, 7.4.4	See section 7.7.8.8 for details.		O	17.023-Value	<See Requirement ID: "Field: Image SVPS Value".>	t-2	
Field: 17.024- Image Quality Score Value	8.17.23, Table 75	<Table 75 lists the value constraints for IQS.>		O	17.024-[QVU, QAV, QAPJ-Value	<See Requirement ID: "Field: Sample Quality Occurrences", "Field: Sample Quality Subfield 1", "Field: Sample Quality Subfield 2'", "Field: Sample Quality Subfield 3".>	t-2	B*
Field: 17.025- Effective Acquisition Spectrum Value	8.17.24, Table 75, Table 76	<Table 75 lists the value constraints for EAS.>	1	O	17.025-Value	{17.025} MO [ASCII(NIR, DEFINED, VIS, RED, UNDEFINED)]		
Field: 17.026-Iris Diameter Value	8.17.25, Table 75, Table 76	<Table 75 lists the value constraints for IRD.>	1	O	17.026-Value	{17.026} MO [10 to 9999] AND MO [integers]		B
Field: 17.027- Specified Spectrum Value	8.17.26	<Table 75 lists the value constraints for SSV.>	1	M ⇑	17.027-LOW-Value	{InfoItem:1 in 17.027} GTE 500 AND {InfoItem:1 in 17.027} MOD 10 EQ 0		B*
			1	M ⇑	17.027-HIG-Value	{InfoItem:1 in 17.027} GTE 510 AND {InfoItem:1 in 17.027} MOD 10 EQ 0		B*
Field: 17.028- Damaged or Missing Eye Value	8.17.27	<Table 75 lists the value constraints for DME.>	1	O	17.028-Value	{17.028} MO ASCII(MA, UC)		B*
Field: 17.029 Reserved	Table 75	Reserved for future use only by ANSI/NIST-ITL.		-	17.029 Reserved			
Field: 17.030- Device Monitoring Mode Value	8.17.28, 7.7.1.3, Table 75, Table 5	See Section 7.7.1.3 for details. <Table 75 lists the value constraints for DMM.>		O	17.030-Value	<See Requirement ID: "Field: Device Monitoring">.	t-2	
Field: 17.031- Subject Acquisition Profile-Iris Value	8.17.29	<Table 75 lists the value constraints for IAP.>		O	17.031-Value	<See Requirement ID: "Field: IAP Values">.	t-2	
Field:	8.17.30	<Table 75 lists the value constraints for	1	O	17.032-	{17.032} MO [1 to 3, 7]		B

Field	Reference	Description	Count	M	Value Name	Value Constraints	
17.032-Iris Storage Format Value		ISF.>			Value		B*
Field: 17.033-Iris Pupil Boundary Value	8.17.31, Table 75, Table 19	<Table 75 lists the value constraints for IPB.>	1	M ⇑	17.033-BYC-Value	{InfoItem:1 in 17.033} MO [ASCII(C,E,P)]	B*
			1	M ⇑	17.033-NOP-Value	{InfoItem:2 in 17.033} MO [2 to 99] AND MO [Integers]	B*
			1	M ⇑	17.033-[HPO, VPO]-Value	For(X EQ 3 to {InfoItem:2 in 17.033}) { IF X MOD 2 EQ 0 {InfoItem:X in Subfield} GTE 0 AND LTE {17.007} AND MO [Integers] ELSE {InfoItem:X in Subfield} GTE 0 AND LTE {17.006} AND MO [Integers] }	B*
Field: 17.034-Iris Sclera Boundary Value	8.17.32, Table 75, Table 19	<Table 75 lists the value constraints for ISB.>	1	M ⇑	17.034-BYC-Value	{InfoItem:1 in 17.034} MO [ASCII(C,E,P)]	B*
			1	M ⇑	17.034-NOP-Value	{InfoItem:2 in 17.034} MO [2 to 99] AND MO [Integers]	B*
			1	M ⇑	17.034-[HPO, VPO]-Value	For(X EQ 3 to {InfoItem:2 in 17.034}) { IF X MOD 2 EQ 0 {InfoItem:X in Subfield} GTE 0 AND LTE {17.007} AND MO [Integers] ELSE {InfoItem:X in Subfield} GTE 0 AND LTE {17.006} AND MO [Integers] }	B*
Field: 17.035-Upper Eyelid Boundary Value	8.17.33, Table 75, Table 19	<Table 75 lists the value constraints for UEB.>	1	M ⇑	17.035-BYC-Value	{InfoItem:1 in 17.035} EQ [ASCII(P)]	B*
			1	M ⇑	17.035-NOP-Value	{InfoItem:2 in 17.035} MO [2 to 99] AND MO [Integers]	B*
			1	M ⇑	17.035-[HPO, VPO]-Value	For(X EQ 3 to {InfoItem:2 in 17.035}) { IF X MOD 2 EQ 0 {InfoItem:X in Subfield} GTE 0 AND LTE {17.007} AND MO [Integers] ELSE {InfoItem:X in Subfield} GTE 0 AND LTE {17.006} AND MO [Integers] }	B*

Field	Ref	Notes		M	Field	Condition	
						AND MO [Integers] }	B*
Field: 17.036- Lower Eyelid Boundary Value	8.17.34, Table 75, Table 19	<Table 75 lists the value constraints for LEB.>	1	M ⇑	17.036- BYC-Value	{InfoItem:1 in 17.036} EQ [ASCII{P}]	B*
			1	M ⇑	17.036- NOP-Value	{InfoItem:2 in 17.036} MO [2 to 99] AND MO [Integers]	B*
			1	M ⇑	17.036- [HPO, VPO]-Value	For{X EQ 3 to {InfoItem:2 in 17.036}) { IF X MOD 2 EQ 0 {InfoItem:X in Subfield} GTE 0 AND LTE {17.007} AND MO [Integers] ELSE {InfoItem:X in Subfield} GTE 0 AND LTE {17.006} AND MO [Integers] }	B*
Field: 17.037-Non- Eyulid Occlusions Value	8.17.35, Table 75, Table 20, Table 21	<Table 75 lists the value constraints for NEO.>	1	M ⇑	17.037- OCY-Value	ForEach(Subfield in 17.037) { {InfoItem:1 in Subfield} MO [ASCII{T,I,L,S}] }	B*
			1	M ⇑	17.037- OCT-Value	ForEach(Subfield in 17.037) { {InfoItem:2 in Subfield} MO [ASCII{L,S,C,R,O}] }	B*
			1	M ⇑	17.037- NOP-Value	ForEach(Subfield in 17.037) { {InfoItem:3 in Subfield} MO [3 to 99] AND MO [Integers] }	B*
			1	M ⇑	17.037- [HPO, VPO]-Value	ForEach(Subfield in 17.037) { For{X EQ 4 to {InfoItem:2 in Subfield}) { IF X MOD 2 EQ 0 {InfoItem:X in Subfield} GTE 0 AND LTE {17.006} AND MO [Integers] ELSE {InfoItem:X in Subfield} GTE 0 AND LTE {17.007} AND MO [Integers] } }	B*

176

Field	Reference	Description	Occ	Field-Value	Value		
Field: 17.038, 17.039 Reserved	Table 75	Reserved for future use only by ANSI/NIST-ITL.	-	17.038,17.039 Reserved	<See Requirement ID: "Field: Type17-CondCode">.	t-2	
Field: 17.040-Range Value	8.17.36, Table 75	<Table 75 lists the value constraints for RAN.>	1	17.040-Value	{17.040} MO [Integers]		B
Field: 17.041-Frontal Gaze Value	8.17.37, Table 75	<Table 75 lists the value constraints for GAZ.>	1	17.041-Value	{17.041} MO [0 to 90] AND MO [Integers]		B
Field: 17.042 to 17.199 Reserved	Table 75	Reserved for future use only by ANSI/NIST-ITL.	-	17.042 to 17.199 Reserved	<See Requirement ID: "Field: Type17-CondCode">.	t-2	
Field: 17.200 to 17.900-User Defined	Table 75	User Defined Fields	-	17.200 to 17.900-User Defined	TRUE		B
Field: 17.901 Reserved	Table 75	Reserved for future use only by ANSI/NIST-ITL.	-	17.901 Reserved	<See Requirement ID: "Field: Type17-CondCode">.	t-2	
Field: 17.902-Annotated Information Value	8.17.39, Table 75, 7.4.1	This is an optional field, listing the operations performed on the original source in order to prepare it for inclusion in a biometric record type. See Section 7.4.1.	O	17.902-[GMT, NAV, OWN, PROJ]-Value	<See Requirement ID: "Field: xx.902-ANN" >.	t-2	
Field: 17.903 to 17.992 Reserved	Table 75	Reserved for future use only by ANSI/NIST-ITL.	-	17.903 to 17.992 Reserved	<See Requirement ID: "Field: Type17-CondCode">.	t-2	
Field: 17.993-Source Agency Name	8.17.40, Table 75	This is an optional field. It may contain up to 125 Unicode characters.	O	17.993-Value	<See Requirement ID: "Field: Source Agency Name" .>	t-2	
Field: 17.994 Reserved	Table 75	Reserved for future use only by ANSI/NIST-ITL.	-	17.994 Reserved	<See Requirement ID: "Field: Type17-CondCode">.	t-2	
Field: 17.995-Associated Context Value	8.17.41, Table 75	See Section 7.3.3	O	17.995-[ACN, ASP]-Value	See Requirement IDs: "Field: xx.995-ASC-ACN" and "Field: xx.995-ASC-ASP">.	t-2	
Field: 17.996-Hash Value	8.17.42, Table 75	See Section 7.5.2	O	17.996-Value	<See Requirement ID: "Field: HAS" >	t-2	
Field: 17.997-Source	8.17.43, Table 75	See Section 7.3.2	O	17.997-[SRN, RSP]-Value	<See Requirement IDs: "Field: xx.997-SOR-SRN" and "Field: xx.997-SOR-RSP">.	t-2	

Representation Value								
Field: 17.998- Geographic Sample Acquisition Location Value	8.17.44, Table 75	See Section 7.7.3		O	17.998-[UTE, LTD,LTM, LTS, LGD, LGM, LGS, ELE, GDC, GCM, GCE, GCN, GRT, OSI, OCV]-Value	<See Requirement IDs: "Field: Geographic", "Field: Geographic", "Field: Geographic-Subfield 1" through "Field: Geographic-Values-SubField 15" >.	t-2	
Field: 17.999- Image Data Valid	8.17.45, Table 75	This field contains the iris image. <The image metadata is checked for validity.>	2	D	17.999- Uncompressed Image Length	IF {17.011} EQ ASCII(NONE) THEN Length(17.999) EQ 17.006} * {17.007}		B
			2	D	17.999- Valid Image Format	IF {17.011} EQ ASCII(JP2) OR ASCII(JP2L) THEN Present(SigBox, HeadBox, ImgBox, EOI-JP2,JP2L) ELSE IF {17.011} EQ ASCII(PNG) THEN Present(PNGSig, IHDR, IDAT, IEND)	t-11	B

Table 6.14 - Assertions for Annex B - Traditional Encoding

Requirement ID	Reference in Base Standard	Requirement Summary	Level	Status	Assertion ID	Test Assertion	Test Note	Implementation Support	Supported Range	Test Result	Applicability
						Annex B: Traditional Encoding					
Traditional-Field: xx.001-Length, First	Annex B	The first field in all records shall contain the length in bytes of the record. For all ASCII or ASCII/Binary records the first field shall also be labeled as field "1".		M	Traditional: xx.001-Length, First	<See Requirement ID: "Field: xx.001-Record Header".>	t-2				
Traditional-Field: xx.002-IDC	Annex B	With the exception of the Type-1 record (See Section 8.1), the second field shall be labeled as field "2" and contain the information designation character / IDC.		M	Traditional: xx.002-IDC	<See Requirement ID: "Field: xx.002-IDC".>	t-2				
Traditional-Record: Type1-7-bit ASCII	Annex B	The data in the Type-1 record shall always be recorded in variable length fields using the 7-bit American Standard Code for Information Interchange (ASCII) as described in ISO/IEC 646. For purposes of compatibility, the eighth (leftmost) bit shall contain a value of zero. All field numbers and information separators shall be recorded in 7-bit ASCII as described in ISO/IEC 646.		M	Traditional: Type-1-ASCII	<See Requirement ID: "Record: Type1-ASCII">	t-2				
Traditional-Field: xx.001, xx.002, xx.999 Ordered	Annex B	Textual fields in Record Types 2 and 9-99 may occur in any order after the first two fields and contain the information as described for that particular numbered field, except for field 999, which shall be the concluding field, when it is included in a record.		M	Traditional: xx.001-First	<See Requirement ID: "Field: xx.001-Record Header".>	t-2				
				M	Traditional: xx.002-Second	<See Requirement ID: "Field: xx.002-IDC".>	t-2				
			1	M	Traditional: xx.999-Last	ForEach(Record in Transaction ST Type(Record) NEQ 4 OR 8 EQ 2 OR GTE 9 AND LTE 99) { IF(Present(Field in Record ST FieldNumber(Field) EQ 999) THEN Field EQ FieldNumber(Last(Field in Record)) EQ 999 }					T

179

Name	Reference	#	M	Type	Description	Test Method / Code	Test ID
Traditional-Transaction: Separators	Annex B, Table 93	1	M	Traditional: Separators FS	In the Type-1, Type-2, Type-9 through Type-99 records, information is delimited by the four ASCII information separators. The delimited information may be items within a field or subfield, fields within a logical record, or multiple occurrences of subfields.	ForEach(Record in Transaction ST Type(Record) MO [1,2,9 to 99]) { Last(Byte in Record) EQ 0x1C }	T
		1	M	Traditional: Separators GS		<Not directly tested. The GS separator is used when parsing fields within a record.>	t-13
		1	M	Traditional: Separators RS,US		<Not directly tested. The RS and US separators are used when parsing subfields and information items.>	t-13
Traditional-Transaction: FS Separator	Annex B	1	M	Traditional: FS Separator	Multiple records within a transaction are separated by the "FS" character, which signals the end of a logical record.	<See Requirement ID: "Traditional-Transaction: Separators" >	t-2
Traditional-Transaction: US, RS, GS, Separators	Annex B	1	M	Traditional: US, RS, GS Separators	The "US" separator shall separate multiple items within a field or subfield; the "RS" separator shall separate multiple subfields, and the "GS" separator shall separate information fields.	<See Requirement ID: "Traditional-Transaction: Separators"; >	t-2
Traditional-Transaction: US Separator Present	Annex B	1	M	Traditional: US Present	In general, if one or more mandatory or optional information items are unavailable for a field or subfield, then the appropriate number of separator characters should be inserted. It is possible to have side-by-side combinations of two or more of the four available separator characters. When data are missing or unavailable for information items, subfields, or fields, there shall be one fewer separator characters present than the number of data items, subfields, or fields required.	<All Information Items, Mandatory and Optional, defined in the standard are separated by the US character—even if the value is missing. All "Presence" and "Count" tests for Information Items first check for the US separator, and then that there is data. For Information Items, presence is defined by the presence of data and the US separator. For Optional Information Items, the US separator is required, but not the data, since their presence is not required>	
Traditional-Transaction: Data Length Minimum	B.1	1	M	Traditional: Data Length Minimum	Each information item, subfield, field, and logical record shall contain one or more bytes of data	<This assertion is tested during the Character Count and Byte Length testing for each Record Type, which will test for 1 byte of data at a minimum. See the following Type-10 Requirement ID's as an example: Field: Type10-CharCount and Field: Type10-SubfieldCharCount .>	t-2
Traditional-Transaction: Big-Endian	B.1.1	2	M	Traditional: Data Big-Endian	Within a file, the order for transmission of both the ASCII and the binary representations of bytes shall be most significant byte first and least significant byte last otherwise referred to as Big-	<Not directly tested. However, the parsing methods use the Big-Endian format when processing transactions.>	t-13

	ID	M	Field	#	Requirement	Assertion	Code
					Endian format. Within a byte, the order of transmission shall be the most significant bit first and the least significant bit last.		
Traditional-Transaction: Date Format	B.1.2	M	Traditional: Date Format	1	Dates shall appear as eight digits in the format YYYYMMDD. The YYYY characters shall represent the year of the transaction; the MM characters shall be the tens and units values of the month; and the DD characters shall be the day in the month.	<This assertion is tested for each field or subfield that requires a date entry. For example, see Requirement ID "Field: 1.005-Local Date Value" >	t-2
Traditional-Field: Agency Code Subfield 2	B.1.3	M	Traditional: Agency Code Subfield 2		The 2007 version of the standard included only agency identifier fields (See Section 7.6). The 2008 added the option of entering an organization name. This capability of the 2008 version is retained in this version of the standard by adding new fields (Field 1.017 Agency names / ANM and Fields xx.993 Source agency name / SAN)	<See Requirement ID: "Field: Agency Codes".>	
Traditional-Transaction: GMT/UTC	B.1.4	M	Traditional: GMT/UTC		GMT/UTC shall be represented as YYYYMMDDHHMMSSZ, a 15-character string that is the concatenation of the date with the time and concludes with the character nal enco YYYY characters shall represent the year of the transaction. The MM characters shall be the tens and units values of the month. The DD characters shall be the tens and units values of the day of the month. The HH characters represent the hour; the MM the minute; and the SS represents the second.	<This assertion is tested for each field or subfield that requires a GMT/UTC entry. For example, see Requirement ID "Field: Geographic-Subfield 1" >	t-2
Traditional-Transaction: Field Numbering	B.1.5	M	Traditional: Field Numbering	1	For the Type-1, Type-2, Type-9 through Type-99 records, each information field that is used shall be numbered in accordance with this standard. The format for each field shall consist of the logical record type number followed by a period rmat for each field followed by a colon al record type number followed by a pepriate to that field. The field number may be any one to nine-digit number occurring between the period field. The colon may be any one to nine-digit number occurring between eld number. This implies that a field number of y one to nine-digit number and shall be	<Not directly tested. However, the parsing methods use the described format when processing transactions.>	t-13

Traditional-Record: Types 1,2,9 ASCII	B.1.5	interpreted in the same manner as a field number of " 2.000000123: Logical Type-1, Type-2, and Type-9 records contain only ASCII textual data fields. The ASCII File Separator "FS" control character (signifying the end of the logical record or transaction) shall follow the last byte of ASCII information and shall be included in the length of the record.	M		Traditional: Type1-ASCII	<See Requirement ID: "Record: Type1-ASCII".>	t-2	T
				1	Traditional: Types 2,9-ASCII	ForEach(Field in Record ST Type(Record) EQ 2 OR 9) { {Bytes(Field)} MO [0x02, 0x03, 0x1C to 0x7E] }		
			M		Traditional: Types 1,2,9-FS Separator	<See Requirement ID: "Traditional-Transaction: Separators".>	t-2	
Traditional-Record: Types 4,8 Binary	B.1.5	The Record Type-4: Grayscale fingerprint image, the Record Type-7: User-defined image record and the Record Type-8: Signature image record contain only binary data recorded as ordered fixed-length binary fields. The entire length of the record shall be recorded in the first four-byte binary field of each record. For these binary records, neither the record number with its period, nor the field identifier number and its following colon, shall be recorded. Furthemore, as all the field lengths of these three records are either fixed and specified, none of the four separato characters ("US", "RS", "GS", or "FS") shall be interpreted as anything other than binary data. For these binary records, the "FS" character shall not be used as a record separator or transaction terminating character.	M		Traditional: Type4-Binary	<See Requirement ID: "Field: Type4-CharType".>	t-2	
			M		Traditional: Type8-Binary	<The test assertions for this type may not be supported in this version of the CTM. If they are supported, they are included under field testing for Record Type-8: Signature image record.>	t-2	
Traditional-Record: Types 10 to 99 Format	B.1.5	The Type-10 through Type-99 records combine ASCII fields with a single binary sample field. Each ASCII field contains a numeric field identifier and its descriptive data. When Field 999 is present in a record it shall appear as the last entry in the record and shall contain the data placed immediately following the colon (er and its descriptive data. When record length field shall contain the length of the record. The ASCII File Separator "FS"control character shall follow the last byte of the compressed or uncompressed	M		Traditional: Types 10 to 99, xx.999 last	<See Requirement ID: "Traditional-Field: xx.001 xx.002 xx.999 Ordered".>	t-2	
			M		Traditional: Types 10 to 99, Record Length	<See Requirement ID: "Field: xx.001-Record Header".>	t-2	
			M		Traditional: Types 10 to 99, FS	<See Requirement ID: "Traditional-Transaction: Separators".>	t-2	

Requirement	Ref	#	M	Traditional	Description	Result	Test
					sample data. The "FS" character shall signify the end of the logical record or transaction and shall be included as part of the record length.		
Traditional-Transaction: Base 64	B.1.5			Traditional: Base 64	The Base-64 encoding scheme (See Annex A: Character encoding information) shall be used for converting non-ASCII text into ASCII form. The field number including the period and colon, for example ll text into ASCII form. The "US", "RS", "GS", and "FS" information separators shall appear in the transaction as 7-bit ASCII characters without conversion to Base-64 encoding.	<Unsupported>	t-4
Traditional-Transaction: Encoding Sets	B.1.6			Traditional: Additional Encoding Sets	In order to effect data and transaction interchanges between non-English speaking or foreign-based agencies, a technique is available to encode information using character encoding sets other than 7-bit ASCII.	<Unsupported>	t-4
Traditional-Field: 1.001 Record Header	B.2.1		M	Traditional: 1.001-Record Length	Field 1.001 Record header shall begin with d header ecode is renamed UTF-16, a record including every character of every field contained in the record and the information separators. The cter of every field contained in the record and theng setsField 1.001 from the next field.	<See Requirement ID "Field: xx.001-Record Header">	t-2
			M	Traditional: 1.001-GS Separator		<See Requirement ID: "Traditional-Transaction: Separators".>	t-2
Traditional-Field: 1.005 Date Format	B.2.1		M	Traditional: 1.005-Date Format	The year, month, and day values in Field 1.005 Date / DAT are concatenated "YYYMMDD"	<See Requirement ID "Field: 1.005-Local Date Value".>	t-2
Traditional-Field: 1.013 DOM	B.2.1		M	Traditional: 1.013-DOM	In Field 1.013 Domain name / DOM, the default is "1.013:NORAM'US'GS"	<Unsupported>	t-3
Traditional-Transaction: FS Separator Replaces GS	B.2.1		M	Traditional: FS Replaces GS	Immediately following the last information item in the Type-1 record (See Section 8.1), an t FS FS n 8.1 following the last information item in the from the next logical record. This "FS" character shall replace the "GS" character that is normally used between information fields. This is the case with all Record Types.	<See Requirement ID: "Traditional-Transaction: Separators".>	t-2
Traditional-Record: Type4-Fields Fixed	B.2.2	2	M	Traditional: Type4 Fields Fixed	The order of fields for Type-4 records is fixed. All fields and data in this record type shall be recorded as binary information.	<Not tested directly, but Type 4 is parsed as unnumbered binary data.>	t-13

Traditional-Record: Type7-Requiremen ts TBD	B.2.3	<This section contains requirements regarding Type-7 records. If this type is supported by the CTM, the requirements will be added.>	-	Traditiona l: Type7 Assertions TBD	<This section contains requirements regarding Type-7 records. If this type is supported by the CTM, the assertions will be added.>	t-15
Traditional-Record: Type8-Requiremen ts TBD	B.2.4	<This section contains requirements regarding Type-8 records. If this type is supported by the CTM, the requirements will be added.>	-	Traditiona l: Type8 Assertions TBD	<This section contains requirements regarding Type-8 records. If this type is supported by the CTM, the assertions will be added.>	t-15
Traditional-Record: Type9-Requiremen ts TBD	B.2.5	<This section contains requirements regarding Type-9 records. If this type is supported by the CTM, the requirements will be added.>	-	Traditiona l: Type9 Assertions TBD	<This section contains requirements regarding Type-9 records. If this type is supported by the CTM, the assertions will be added.>	t-15
Traditional-Transaction: Type11-Reserved	B.2.7	This Record Type is reserved for future use as Voice data.	M	Traditiona l: Type11-Reserved	<See Requirement ID: "Transaction: Reserved Records".>	t-2
Traditional-Transaction: Type12-Reserved	B.2.8	This Record Type is reserved for future use as Dental data.	M	Traditiona l: Type12-Reserved	<See Requirement ID: "Transaction: Reserved Records".>	t-2
Traditional-Field: 13.014 RS Separated	B.2.9	For Field 13.014: Search position descriptors / SPD, multiple portions of the EJI may be listed and separated by the "RS" character.	M	Traditiona l: 13.014-RS Separated	<See Requirement ID: "Field: SPD PPD Values.>	t-2
Traditional-Field: 13.015 Subfields	B.2.9	For Field Field 13.015: Print position coordinates / PPC, the six information items within the field are separated by five s US PPCD Dental data.SUblock shall belowed by tedal definitions may be repeated as subfields separated by the dRSblock shall	M	Traditiona l: 13.015-Subfields	<See Requirement ID: "Field: PPC-Subfield Occurrences" >	t-2
Traditional-Field: 13.024 Subfields	B.2.9	Field Field 13.024: Latent quality metric/ LQM may contain one or more subfields, each consisting of four information items separated by the" US" character. The subfield may be repeated for each latent image and quality algorithm used, separated by the "RS" character.	M	Traditiona l: 13.024-Subfields	<See Requirement IDs: "Field: Sample Quality Occurrences" to "Field: Sample Quality Additional Subfield".>	t-2

6.3 Test notes

The following test notes provide clarification of the assertion text provided in the Test Assertion column. The test notes contain various types of information including:

- Additional information to help clarify complex assertions such as image metadata and IDC comparisons.
- Explanations of decisions made when the base standard is not clear or contains possible discrepancies.
- "Exception" which refers to any AN-2011 requirement that does not have an associated assertion defined in this document.

t-1. Semantic (Level 3) assertions are not included in the tables except to intentionally clarify that the assertion is L3. An explanation may be given regarding the assertion's categorization as L3.

t-2. The assertions for this requirement are listed in another section of the table as described in the "Test Assertion" column.

t-3. Assertions related to Domain Names and Application Profile Specifications are not addressed in the tables.

t-4. Assertions related to Character Sets other than 7-bit ASCII or binary are not addressed in the tables.

t-5. For fields with matching IDC's, refer to the tables below to determine the test result based upon the biometric sample types being compared. For matching IDC's that belong to record types not listed below, the test will only check that the record types are the same.

Table 6-15 - IDC ID Location Comparison

IDC: ID Location Comparison		
Record Type	Biometric	Field For Comparison ID
4	FINGER	{Byte:1 in 4.004}
10	{10.003} <SCAR, MARK, TATTOO, or FACE>	IF {10.003} MO [ASCII(SCAR,MARK, TATTOO)] THEN {InfoItem:1 in 10.040}
13	FINGER	{InfoItem:1 in 13.013}
14	FINGER	{InfoItem:1 in 14.013}
15	PALM	{15.013}
17	IRIS	{17.003}
19	PLANTAR	{19.013}

Table 6-16 - IDC Comparison Results

	IDC: Comparison Results		
First Record	Second Record	Comparison IDs	Result
FINGER	FINGER	Same	Ok
FINGER	FINGER	Different	Error
FINGER	NEQ FINGER	NA	Error
FACE	FACE	NA	Ok
FACE	NEQ FACE	NA	Error
IRIS	IRIS	Same OR Either EQ 0	Ok
IRIS	IRIS	Different AND Neither EQ 0	Error
IRIS	NEQ IRIS	NA	Error
SCAR	SCAR	Same	Ok
SCAR	SCAR	Different	Warning
SCAR	MARK Or TATTOO	Same or Different	Warning
SCAR	FINGER, FACE Or IRIS	NA	Error
MARK	MARK	Same	Ok
MARK	MARK	Different	Warning
MARK	SCAR Or TATTOO	Same or Different	Warning
MARK	FINGER, FACE Or IRIS	NA	Error
TATTOO	TATTOO	Same	Ok
TATTOO	TATTOO	Different	Warning
TATTOO	SCAR Or MARK	Same or Different	Warning
TATTOO	FINGER, FACE Or IRIS	NA	Error
PALM	PALM	Same	Ok
PALM	PALM	Different	Error
PALM	NEQ PALM	NA	Error
PLANTAR	PLANTAR	Same	Ok
PLANTAR	PLANTAR	Different	Error
PLANTAR	NEQ PLANTAR	NA	Error

t-6. UTC has replaced GMT. Date and time are defined in section 7.7.2 of the standard. The set of values ValidUTC/GMT is described in section 7.7.2.2 of the standard and is always less than the current date and time. ValidUTC/GMT is in the form YYYYMMDDHHMMSSZ; NIEM-ValidUTC/GMT is in the form YYYY-MM-DDThh:mm:ssZ. The ValidLocalDate is in the form YYYYMMDD; NIEM-ValidLocalDate is in the form YYYY-MM-DD.

t-7. Assertions for alternate coordinate systems are not included in the tables.

t-8. Refer to http://earth-info.nga.mil/GandG/coordsys/grids/utm.html to determine valid values for the band of latitude and grid zone.

t-9. L2 and L3 assertions associated with SAP, FAP, and IAP are not included in the tables. Some of the assertions, such as determining the conditions under which the samples were collected to ensure the SAP, FAP, or IAP levels, are not feasible to test.

t-10. IBIA Vendor Registry is a registry that maps the QAV value to a registered CBEFF Biometric Organization. Since the standard does not require that the value be registered with IBIA, the test assertions will accept any values for QAV.

t-11. All assertions associated with compressed image types used the image metadata and not the image data itself. Metadata features from each image type are defined in the "Image Metadata" table below. Combining the "Term" with the "Image Type" provides the specific implementation. For example, {Image Height-PNG} is equivalent to '2nd parameter of the IHDR chunk". Note that for NIEM encoding, the image data must first be converted from Base-64.

Image Metadata		
Term	Image Type(s)	Implementation
Image Width	JPEG, JPEGL	4th parameter of the Frame Header not counting the SOF marker
	JP2, JP2L	2nd parameter of Image Header box
	PNG	1st parameter of IHDR chunk
	WSQ	5th parameter of SOF not counting the SOF marker
Image Height	JPEG, JPEGL	3rd parameter of the Frame Header not counting the SOF marker
	JP2, JP2L	1st parameter of Image Header box
	PNG	2nd parameter of IHDR chunk
	WSQ	4th parameter of SOF not counting the SOF marker

Table 6-17 - Image Metadata

Field	Format	Description
Sampling Units	JPEG, JPEGL	4th parameter in JFIF Header not counting the APP0 Marker
	JP2, JP2L	Undefined
	PNG	3rd parameter of PHYS chunk
	WSQ	Undefined
Horizontal Density	JPEG, JPEGL	5th parameter in JFIF Header not counting the APP0 Marker
	JP2, JP2L	Undefined
	PNG	1st parameter in PHYS Chunk
	WSQ	Undefined
Vertical Density	JPEG, JPEGL	6th parameter in JFIF Header not counting the APP0 Marker
	JP2, JP2L	Undefined
	PNG	2nd parameter in PHYS Chunk
	WSQ	Undefined
BPX	JPEG, JPEGL	2nd parameter of the Frame Header not counting the SOF marker
	JP2, JP2L	7 LSB of 4th parameter of ImgBox + 1 if 4th parameter of ImgBox is not 255
	PNG	3rd parameter of IHDR chunk
	WSQ	Undefined
CSP	JPEG, JPEGL	Undefined
	JP2, JP2L	4th parameter of Colour Specification box
	PNG	4th parameter of IHDR chunk
	WSQ	Undefined
Encoder Version	WSQ	10th parameter of SOF not counting the SOF marker
SOI	JPEG, JPEGL	Start of JPEG type image.
	WSQ	Start of WSQ image.
SOF	JPEG, JPEGL	Start of frame in a JPEG type image.
	WSQ	Start of WSQ image.
EOI	JPEG, JPEGL	End of a JPEG image.
	JP2, JP2L	End of JP2 image.
	WSQ	End of WSQ image.
SOB	WSQ	Start of block in a WSQ image.
SigBox	JP2,JP2L	Signature Box that marks the start of a JP2 type image.
HeadBox	JP2,JP2L	Header Box in a JP2 type image.
ImgBox	JP2, JP2L	Image Header Box in a JP2 type image.
PNGSig	PNG	Signature of a PNG image.
IHDR	PNG	Image Header Chunk in a PNG image.
IDAT	PNG	Image Data Chunk in a PNG image.
IEND	PNG	Image End Chunk in a PNG image.
JFIF Header	JPEG, JPEGL	Frame for specifying JPEG type image metadata. Its inclusion is required by the standard.

| PHYS Chunk | PNG | An optional Chunk in a PNG image that may be used if present to verify image attributes. |

t-12. Some image formats (such as PNG) use scale units other than pixels per inch (ppi) or pixels per centimeter (ppc), which are the measurements used in the requirements of the base standards. For those image formats, the use of a conversion factor is necessary to convert the pixel scale to the correct units (either ppi or ppc). This conversion may result in a decimal value that cannot be held in the THPS or TVPS fields, which hold Integer values only. In all such cases, the AN-2011 standard does not state which method should be used to round the result. The CTM suggests using both the mathematical floor and ceiling functions of the converted value as acceptable values for comparison to the field value to account for various rounding methods.

t-13. The assertion is addressed during parsing.

t-14. Further research is needed to determine the feasibility of testing for ASEG requirements related to the polygon structure.

t-15. These requirements are related to record types currently not supported.

t-16. Only the minimum requirement of WSQ 2.0 is tested, because determining the platen size used is a level-3 assertion. A value of 1 in the Ev field of the WSQ metadata represents WSQ 2.0, while a value of 2 represents WSQ 3.1.

t-17. The NCIC codes are available at http://oregon.gov/OSP/CJIS/NCIC.shtml

6.4 Test assertions exceptions table

An "exception" refers to any AN-2011 requirement that does not have an associated assertion defined in this document. Table C.1 identifies and provides justification for all exceptions present in the tables.

Table 6.18 - Exceptions Table

Exception	Section	Requirement Summary	Justification
Domain Names / Application Profile Specifications	5.3.2	Data contained in this record shall conform in format and content to the specifications of the domain name(s) as listed in Field 1.013 Domain name / DOM found in the Type-1 record, if that field is in the transaction. The default domain is NORAM. Field 1.016 Application profile specifications / APS allows the user to indicate conformance to multiple specifications. If Field 1.016 is specified, the Type-2 record must conform to each of the application profiles. A DOM or APS reference uniquely identifies data contents and formats. Each domain and application profile shall have a point of contact responsible for maintaining this list. The contact shall serve as a registrar and maintain a repository including documentation for all of its common and user-specific Type-2	The format and content of the record are defined by the DOM or APS. Each DOM and APS has related record-content definitions that may be updated. The evolving nature of the DOM and APS definitions and nature of using registrars makes testing for conformance via the CTS very difficult.

	6	data fields. As additional fields are required by specific agencies for their own applications, new fields and definitions may be registered and reserved to have a specific meaning. When this occurs, the domain or application profile registrar is responsible for registering a single definition for each number used by different members of the domain or application profile . An implementation domain, coded in Field 1.013 Domain name / DOM of a Type-1 record as an optional field, is a group of agencies or organizations that have agreed to use preassigned data fields with specific meanings (typically in Record Type-2) for exchanging information unique to their installations. The implementation domain is usually understood to be the primary application profile of the standard. New to this version of the standard, Field 1.016 Application profile specifications / APS allows multiple application profiles to be referenced. The organization responsible for the profile, the profile name and its version are all mandatory for each application profile specified. A transaction must conform to each profile that is included in this field. It is possible to use Field 1.016 and / or Field 1.013. A specified implementation domain and specified application profiles must all have the same definition for fields, subfields and information items that are contained in the transaction.	Since the "transaction must conform to each profile" included in the field, and those profiles are defined by the listed agency, the CTS would have to retrieve the latest requirements from the agency. Also, testing that all specified DOM and APS have the same definitions for fields, subfields, and information items is not feasible.
Alternate Character Sets	5.6, Table 2	Field 1.015 Character encoding/DCS is an optional field that allows the user to specify an alternate character encoding... Field 1.015 Character encoding/DCS contains three information items: the character encoding set index/ CSI, the character encoding sent name/CSN, and the character encoding set version/CSV. The first two items are selected from the appropriate columns of Table 2.	Table 2 lists ASCII, UTF-16, UTF-8, and UTF-32 as possible encodings. However, the table also allows "User-defined" character encoding sets. Testing for conformance to user-defined character encoding sets is not feasible for the CTS. Further research is needed to support Character sets other than 7-bit ASCII.
Alternate Coordinate System	7.7.3, Table 4	The ninth information item is the geodetic datum code / GDC10. It is an alphanumeric value of 3 to 6 characters in length. This information item is used to indicate which coordinate system was used to represent the values in information items 2 through 7. If no entry is made in this information item, then the basis for the values entered in the first eight information items shall be WGS84, the code	Table 4 lists 22 coordinate systems and the option to include "Other" types as well. It is not feasible for the CTS to test conformance to all coordinate

		for the *World Geodetic Survey 1984 version - WGS 84 (G873)*. See Table 4 for values.	systems, specifically those that are listed by the user under "Other".
	7.7.3	A fourteenth optional information item geographic coordinate other system identifier / OSI allows for other coordinate systems. This information items specifies the system identifier. It is up to 10 characters in length. Examples are: • MGRS (Military Grid Reference System) • USNG (United States National Grid) • GARS (Global Area Reference System) • GEOREF (World Geographic Reference) • LANDMARK (e.g. hydrant) and position relative to the landmark. A fifteenth optional information item, is the geographic coordinate other system value / OCV. It shall only be present if OSI is present in the record. It can be up to 126 characters in length. If OSI is LANDMARK, OCV is free text and may be up to 126 characters. For details on the formatting of OCV for the other coordinate systems shown in OSI as examples, see http://earth-info.nga.mil/GandG/coordsys/grids/referencesys.html	While some examples are listed (MGRS, USNG, GARS, GEOREF, LANDMARK), there may be others that are not listed. It is not feasible for the CTS to test conformance to these coordinate systems, specifically those that may be included but are not listed as examples.
Subject Acquisition Profiles SAP/FAP/IAP	7.7.5, Table 8, Table 9, Table 10	A subject acquisition profile is used to describe a set of characteristics concerning the capture of the biometric sample. These profiles have mnemonics SAP for face, FAP for fingerprints and IAP for iris records.	It is not feasible to test if the image was captured under the conditions specified by the SAP, FAP or IAP level as defined in Tables 8, 9 and 10. However, the fields will be tested for valid level values.
Open and Closed Paths	7.8	Several Record Types define open paths (also called contours or polylines) and / or closed paths (polygons) on an image. They are comprised of a set of vertices. For each, the order of the vertices shall be in their consecutive order along the length of the path, either clockwise or counterclockwise. (A straight line of only two points may start at either end.) A path may not have any sides crossing. No two vertices shall occupy the same position. There may be up to 99 vertices. An open path is a series of connected line segments that do not close or overlap. A closed path (polygon) completes a circuit. The closed path side defined by the last vertex and the first vertex shall complete the polygon. A polygon shall have at least 3 vertices. The contours in Record Type-17: Iris image record can be a circle	Further research is needed to determine the feasibility of testing for: -simple, plane figure -no sides crossing -no interior holes

or ellipse. A circle only requires 2 points to define it (See Table 16). There are two different approaches to the paths in this standard. The 2007 and 2008 version of the standard used paths for Field 14.025: Alternate finger segment position(s) / ASEG.

That approach has been retained in this version for all paths except in the Extended Feature Set (EFS) of Record Type-9. The EFS adopted an approach expressing the path in a single information item, which is different than that used in other record types.

192

Annex A: Support for AN-2011 Record Types and Interrelated Field

A.1 Support for Record Types

The following table outlines the conformance testing support for each of the record types defined by the AN-2011 standard.

Table A.1 - AN-2011 Record Type Support

	Support for AN-2011 Record Types	
Number	**Record Contents**	**Support**
1	Transaction Information	Full Support. See Record Type-1: Transaction information record.
2	User-defined descriptive text	None.
3	Low-resolution grayscale fingerprint image (Deprecated)	Deprecated requirement only. See Record Type-3: DEPRECATED.
4	High-resolution grayscale fingerprint image	Full Support. See Record Type-4: Grayscale fingerprint image.
5	Low-resolution binary fingerprint image (Deprecated)	Deprecated requirement only. See Record Type-5: DEPRECATED.
6	High-resolution binary fingerprint image (Deprecated)	Deprecated requirement only. See Record Type-6: DEPRECATED.
7	User-defined image	None.
8	Signature image	None.
9	Minutiae data	None.
10	Face, other body part, or scar, mark tattoo (SMT) image	Full Support. See Record Type-10: Facial, other body part and SMT image record.
11	Voice Data (future addition to the standard)	Reserved requirement only. See Reserved Record Types.
12	Dental record data (future addition to the standard)	Reserved requirement only. See Reserved Record Types.
13	Variable-resolution latent friction ridge image	Full Support. See Record Type-13: Friction-ridge latent image record.
14	Variable-resolution fingerprint image	Full Support. See Record Type-14: Fingerprint image record.
15	Variable-resolution palmprint image	Full Support. See Record Type-15: Palm print image record.
16	User-defined variable-resolution testing image	None.
17	Iris Image	Full Support. See Record Type-17: Iris image record.
18	DNA data	None.
19	Variable-resolution plantar image	None.
20	Source representation	None.
21	Associated context	None.
22-97	Reserved for future use	Reserved requirement only. See Reserved Record Types.
98	Information assurance	None.
99	CBEFF biometric data record	None.

It is necessary to provide support for some fields contained in unsupported record types because of structural requirements and relationships to supported record types. These fields are outlined in the table below.

A2 Interrelated Field Support

Table A.2 - AN-2011 Interrelated Field Support

	Support for AN-2011 Interrelated Fields	
Number	**Field Contents**	**Support**
xx.001	Record header	All Record Types. See Field: xx.001-Record Header
xx.002	Information designation character / IDC	All Record Types except Record Type-1. See Field: xx.002-IDC
xx.995	Associated Context / ASC	Record Types 10 and above, not including 21 and 98. See Field: xx.995-ASC through Field: xx.995-ASC-ASP
xx.997	Source Representation / SOR	Record Types 10 and above, not including 18, 21, and 98. See Field: xx.997-SOR through Field: xx.997-SOR-RSP
xx.016	Segments / SEG	Record Types 20 and 21. See Field: xx.997-SOR-RSP and Field: xx.995-ASC-ASP
xx.021	SRN, ACN	Record Types 20 and 21. See Field: xx.997-SOR-SRN and Field: xx.995-ASC-ACN

Acknowledgements

The Information Technology Laboratory (ITL) of the National Institute of Standards and Technology (NIST) sponsored the development of this conformance testing methodology (CTM) for AN-2011, Data Format for the Interchange of Fingerprint, Facial & Other Biometric Information under the Conformance Testing Methodology Working Group (WG). This WG is chaired by Elham Tabassi of the NIST/ITL Information Access Division (IAD). Membership in the Working Group that developed this testing methodology is from NIST, other US Government agencies, and industry.

NIST/ITL CSD's staff Fernando L. Podio, Dylan Yaga, and Christofer J. McGinnis who were the editors of this publication and developed an initial version of an ANSI/NIST-ITL 1-2011 conformance test tool have contributed the test assertions implemented in that test tool for use in this publication. This work was sponsored, in part, by the Department of Homeland Security/US-VISIT Program.

The following individuals participated in the Conformance Testing Methodology Working Group and contributed to the development of this testing methodology:

Name	Organization
Anne Wang	3M Cogent
Scott Hills	Aware
Chris White	Booz Allen Hamilton
Matthew Young	Booz Allen Hamilton
Kim Woods	CTR Biometric Image Management Agency
William R. Graves	DHS
Ryan L. Triplett	DoD
Jennifer Stathakis	FBI
B. Swann	FBI
Charles Schaeffer	Florida Dept. of Law Enforcement (FDLE)

Christofer J. McGinnis	ID Technology Partners
Mike McCabe	ID Technology Partners
Kaustubh Deshpande	L1
Adam Rosefsky	MaxVision
Michael D. Garris	NIST
Patrick J. Grother	NIST
Michael D. Hogan	NIST
Kevin C. Mangold	NIST
Fernando L. Podio	NIST
Elham Tabassi	NIST
Bradford Wing	NIST
Dylan Yaga	NIST
Brian Finegold	Noblis
Austin Hicklin	Noblis
John Mayer-Splain	Noblis
Bonny Scheier	SABER
C. J. Lee	SAIC
Sudhi Umarji	Trusted Federal
Cindy Wengert	Trusted Federal

www.ingramcontent.com/pod-product-compliance
Lightning Source LLC
Chambersburg PA
CBHW081443170526
45166CB00008B/2293

* 9 781500 177539 *